SOLUTIONS MANUAL TO ACCOMPANY

BRADY / HUMISTON / THIRD EDITION

GENERAL CHEMISTRY

PRINCIPLES AND STRUCTURE

THEODORE W. SOTTERY
University of Southern Maine

1807 175 YEARS OF PUBLISHING 1982

JOHN WILEY & SONS NEW YORK CHICHESTER BRISBANE TORONTO SINGAPORE

ISBN 0 471 09964 3

Printed in the United States of America

10 9 8

DEDICATED

IN MEMORIAM

C. THEODORE SOTTERY

July 30, 1896 + October 30, 1981

The finest man

The greatest teacher

With the sweetest smile

That I have ever known.

PREFACE

This Problem Solutions Manual has been prepared as a suppliment to GENERAL CHEMISTRY - principles and structure - 3rd edition by James E. Brady and Gerard E. Humiston. That text provides an excellent introduction to the theories and principles of modern chemistry. The knowledge and understanding of these principles is the foundation of proficiency in chemistry, but the ability to repeat the words and phrases of the text or even restate the principles in your own words does not signify mastery of chemistry. You must be able to utilize your knowledge to deduce the correct answers to questions about chemical systems and make logical predictions about the results of processes which you have not observed directly. The comprehensive review questions which are included with each chapter of the text are designed to help you to develop this proficiency.

Chemistry is a quantitative science and the development of the skills required for the solution of chemical problems is a vital portion of your study of chemistry. For many students this is the most difficult and confusing part of chemistry! To aid you in the transition from understanding principles to solving problems the authors have provided detailed examples of many types of chemistry problems as an integral part of each chapter of the text. These worked out problems are designed to clarify and explain each step of the process which starts with the initial data and and employs logic and the techniques of mathematics to determine the numerical answer. When you understand the procedures used in a worked out example you are ready to take the next step, the solution of a similar problem on your own. If you can complete the solution without any help, (you are great!) try a few of the same type to be sure and then go on to another kind. BUT, if you can't make it all the way to the answer on the first try, cheer up, YOU HAVE ALOT OF COMPANY! (The author of this solutions manual, for one!) That's right, as a freshman chemistry student this author was often stumped. (No, I did not write on clay tablets in cuneiform!) This manual was designed to help those who have difficulty with chemistry problems. I hope all you 'brains' who can zip right through all the problems will take a casual peek at these solutions (to see if your method is superior to mine?) but this manual was not designed for the 'brains'.

There is a detailed solution in this manual for each of the review problems in the text. These solutions are compatible with the worked out examples, but you shouldn't expect them to have exactly the same format. As a discussion of the technique of writing poetry is not a poem, so a worked out example in the text is not identical to the format which an experienced chemist uses to actually solve a problem.

In general the solutions in this manual are designed as models for you to use as you evolve your own individual style, but some explanations and clarifications are inserted to further your knowledge. For example, numbers which express the results of measurements or those which result from calculations which used measured quantities always involve some uncertainty. The degree of this uncertainty is indicated by the number of "significant figures" (usually abbreviated as s.f. in this manual) to which the number is expressed. The data number with the fewest significant figures (largest uncertainty) limits the number of significant figures to which the answer can be expressed, just as it is the <u>weakest</u> link which determines the strength of the chain. To help you become familiar with this principle, the "weakest link" number will often be indicated by the symbol • as in the example: 6.022•. Some of the numbers used in problem calculations are not the results of measurements and do not have uncertainty associated with them. Such numbers will never represent the "weakest link" in a problem solution and will not set the limit on the number of significant figures of the answer. Some examples are DEFINED NUMBERS like: 1 foot = 12 inches, SPECIFIED NUMBERS as in, "ten grams of oxygen contain ? atoms" (if the specification were 10.0g instead of ten grams it is clear that a measurement number, good to three s.f. should be assumed) and CARDINAL NUMBERS (counting numbers) like the number of hydrogen atoms in a molecule of water (H_2O) or the number of seats in the lecture hall. To remind you which numbers in a problem are 'exact', the symbol * will be used as in the example 12*inches. (These symbols would not be included in routine problem solutions and you should not use them when you work problems.

There are many "tricks of the trade" associated with techniques of problem solving and I will try to pass some of the ones which I have found useful along to you. Don't be alarmed if your answer is slightly different from the one given here. A slight difference might be due to rounding numbers at a different stage of the calculation process. In this manual only final answers (double underlined) are always rounded to the correct number of significant figures, intermediate answers may or may not be rounded off. (A good general rule is to carry one more place throughout the calculation than is allowed for the final answer.) All of the answers in this manual have been checked independently but the probability that every error has been eliminated is not high! If you note an error please let me know. I would be very grateful for your help.

I wish you the best of luck in this and future chemistry courses. If this manual helps you over some of the rough spots along the road to your mastery of problem solving skills, I will be very pleased!

Theodore W. Sottery
University of Southern Maine
Portland, Maine

CONTENTS*

An Introduction to This Problem Solutions Manual		1
Chapter 1	INTRODUCTION	11
Chapter 2	STOICHIOMETRY: chemical arithmetic	23
Chapter 3	ATOMIC STRUCTURE AND THE PERIODIC TABLE	50
Chapter 4	CHEMICAL BONDING: general concepts	56
Chapter 6	CHEMICAL REACTIONS IN AQUEOUS SOLUTION	58
Chapter 7	GASES	79
Chapter 8	STATES OF MATTER AND INTERMOLECULAR	98
	FORCES	98
Chapter 10	PROPERTIES OF SOLUTIONS	105
Chapter 11	CHEMICAL THERMODYNAMICS	120
Chapter 12	CHEMICAL KINETICS	136
Chapter 13	CHEMICAL EQUILIBRIUM	145
Chapter 15	ACID-BASE EQUILIBRIA IN AQUEOUS SOLUTION	159
Chapter 16	SOLUBILITY AND COMPLEX ION EQUILIBRIA	194
Chapter 17	ELECTROCHEMISTRY	213
Chapter 18	CHEMICAL PROPERTIES OF THE REPRESENTATIVE METALS	228
Chapter 19	THE CHEMISTRY OF SELECTED NONMETALS	231
	PART I	231
Chapter 21	THE TRANSITION ELEMENTS	232
Chapter 24	NUCLEAR CHEMISTRY	233

*Chapters 5, 9, 14, 20, 22, and 23 do not include REVIEW PROBLEMS.

INTRODUCTION

TO THE STUDENT: If you have not read the preface of this manual, please do so before you continue. Some of the material covered in the preface is vital to your successful use of this manual. There are two topics which are fundamental to the problem solutions given in this manual. These are SIGNIFICANT FIGURES and DIMENSIONAL ANALYSIS and both will be treated here before we begin to solve specific numerical problems.

SIGNIFICANT FIGURES

Chemistry problems often involve numerical values which express the results of measurements. These values are not exact since there is always uncertainty associated with any measurement. (The magnitude of this uncertainty depends upon the device(s) used to make the measurement and the skill and care of the investigator.)

The number of figures used to express a measurement number indicates the precision of the measurement, with the final number implying both some knowledge and some uncertainty. Suppose, for example, you were asked to measure the length of a room with a meter stick which had been sandpapered to remove all of the graduation marks. If we assume that the room is more than one meter in length, you would have to move the stick, keeping track of the number of times you picked it up and set it down again. Finally, however, the far wall would be too close (less than one meter) and when you placed the end of the stick against that wall, a portion would project back into the section already measured. How should you record the result of this measurement? If you used the stick nine times before the remaining distance was less than one meter, you might record the length as '9 meters' (9m). But since you know the length was more than 9 meters; perhaps you should record 10m. Either of those values, however, conveys less information than you received when you looked at the position of the stick that last time. You can estimate by eye what portion of the stick was required to fill the remaining distance. Since the metric system uses increments of ten, if slightly more than half of the stick is required, you might estimate six tenths and record the room length as 9.6m. This value would indicate your opinion that the portion of the stick needed to fill the final section was more than 1/2 but less than 2/3. (More than 0.5 but less than 0.7 is exactly what you imply when you express the room length as 9.6m.) NOTE: Just as we stated before, that final number (6) conveys both the knowledge that the length is more than 9m and less than 10m and the uncertainty about the number of tenths of a meter. To be more explicit you might express your opinion that the room is > 9.5m but < 9.7m by writing: 9.6m ± 0.1m. Although research results are often shown with this explicit notation, for general use we will assume that the number 9.6m implies an uncertainty of at least ± 0.1m. We describe '9.6' as having "two significant figures". (The '6' is certainly significant since it conveys the information that the measured distance is a little more than nine and one-half meters.)

If you carefully divided the stick into ten equal portions, with numbered graduation marks, and repeated the measurement, you would be able to see that the final distance was between say the six tenths mark and the seven tenths mark and you could then estimate (in tenths of course!) how far between the marks. This more refined measuring device (a meter stick with tenths graduations) would allow a more precise measurement of the room length, say 9.63m, and the use of three "significant figures" would properly indicate that the uncertainty has been reduced to about ± 0.01m (1cm). A common meter stick is actually graduated to thousandths of a meter (millimeters) and it would seem that your use of such a meter stick would allow you to express the room length to the exact number of millimeters and even estimate tenths of a millimeter. NO! You would be telling a "scientific lie" about the precision of your measurement. WHY? Everything was going so well, can't we continue? Each time you picked up the stick and set it down again you introduced some uncertainty and after nine repetitions this uncertainty would certainly be greater than one millimeter. The use of this "better" meter stick would not give you any more precise knowledge of the room length! A ten meter measuring tape would help by eliminating the "pick it up and set it down" error but the tape would expand and contract with temperature changes (The room would expand and contract also but not to the same extent as the tape.) and careful measurements on different days would not be likely to agree to the nearest millimeter!

You can't escape uncertainty when you make measurements but you can reduce it to a minimum and express your results so that the magnitude of the uncertainty is indicated. It is desirable to express other data numbers used in a problem (and intermediate answers) to one more significant figure (if possible) than the least precise data number. The final answer, however, <u>must</u> be limited to the correct number of significant figures. For multiplication and division that number is determined by the data value with the smallest number of significant figures. For addition and subtraction the data number with the fewest decimal places determines the number of decimal places in the answer. These two rules are justified by the example which follows: Consider the following numbers- 351.8 (4 s.f.) and 5.64 (3 s.f.) A calculator gives the product 1984.152 which, by the multiplication rule, should be rounded to 3 s.f. (1980). Since we must assume an uncertainty of at least ± 1 in the last place, the values may be as large as 351.<u>9</u> and 5.6<u>5</u> and the product of these two would be the largest value which we might expect for an answer (1980.235 by calculator) which, rounded to three significant figures would be <u>1990</u>. On the other hand the numbers might be as small as 351.<u>7</u> and 5.6<u>3</u> so the smallest expected product

(1980.071) would round to 1980. Since we obtained different numbers in the third place it is clear that there is some uncertainty in that place and it is indeed the last ''significant figure''. This is exactly what the use of the multiplication rule predicted. Demonstrate the division rule for yourself. Divide 351.8 by 5.64 and round the answer by the division rule. Then divide 351.9 by 5.63 to obtain the largest quotient and 351.7 by 5.65 to calculate the smallest quotient. By comparing the values determine the place at which the answer is uncertain. Did the division rule work?

THE SUM OF THE TWO NUMBERS:	LARGER SUM:	SMALLER:
351.8	351.9	351.7
5.64	5.65	5.63
357.44	357.55	357.33

NOTE: The three sums show variation in the 1st decimal place, so that place is the last s.f. The addition rule predicts that the answer should be expressed to one decimal place.

(USE A SIMILAR TECHNIQUE TO DEMONSTRATE THAT THE SUBTRACTION RULE ALSO YIELDS THE CORRECT NUMBER OF SIGNIFICANT FIGURES.)

Significant figures are not a fussy detail designed to make your problem solving activities more dificult. Significant figures are vital if you wish to avoid telling "scientific lies"!

DIMENSIONAL ANALYSIS

Dimensional analysis is a general problem solving technique which considers the units (names) of the quantities involved in the problem and uses the interrelationships amoung those units to solve the problem. (The "factor label method", "units conversion", the "units method" and other terms are alternative names for the same process.) As you progress through the chapters of this manual, you will encounter applications of this technique to a wide variety of chemistry problems. The application of dimensional analysis, however, is not limited to chemistry problems; the procedure works just as well in fields like physics or engineering. So long as the quantities considered in the problem have "names" (units) this system can be utilized.

What is the only quantity which you can use as a multiplier times "X" and always have the answer equal to "Y"? That may be a real mindblower if you are thinking of numbers like 5, 1/2, or 3.14159 since using any of these would give an equation which is only valid for certain sets of X,Y values. The answer we are looking for is ridiculously simple, once you think of it, it is Y divided by X! The equation: $X \cdot Y/X = Y$ is always true for any values of X and Y because X cancels X leaving the identity Y = Y. (Someone is crying, "foul!" The rest are grumbling that you didn't dream that I meant that silly answer! Hang in there; I don't want to turn you off, I just want you to see that the basic technique of dimensional analysis is very simple and straightforward; you just treat the units in the problem like X, Y and Z in algebra.

Although you may not have thought in these terms before, the answer to a problem has to be equivalent to the quantity which you used to start the problem. If you want to know how many millimeters there are in 25 feet, the answer in mm must represent the same distance as 25 feet. Working a problem involves expressing the quantity given in the problem data in a different form. Do you see the implication? The process of solving a problem must not change the fundamental starting quantity, and the only multiplier which doesn't change that starting quantity is 1! So those Y/X units ratio multipliers which you use must have numbers along with the units that make the complete ratio equal to 1. Since 12 inches = 1 foot, you can divide both sides of the equality by 1 foor and obtain 12 inches / 1 foot = 1 or divide both sides by 12 inches yielding 1 foot / 12 inches = 1. Since both of these units ratios equal one (we will refer to them as "unity ratios") either could be used a a multiplier in any problem solution with complete confidence that we would not be changing the fundamental quantity we used to start the problem. (Tar and feathers for anyone who uses 13 inches / 1 foot as a multiplier in a problem solution! That ratio doesn't equal 1 and the answer will not be equivalent to the starting quantity. A big red "X" and no credit for that one!) Whenever the numerator and the denominator of a units ratio are equal (or equivalent) we have a "unity ratio" which can be used as a multiplier whenever we need it.

To sum up, the technique of dimensional analysis may be regarded as a series of steps. First, you locate the answer requested by the problem (look for key words like "calculate - -", "how much - - -", "how many - -", "what is the - -", "what would be - -", "determine the - -", etc. to locate the "answer".) (There are other words such as "which - -", "compare - - ", "what are the - ", which alert us that we may have to calculate more than a single answer.) When you have located the answer you must supply units to express that answer so that you have a clear target to direct your problem solving effort. Second, you select the specific quantity from the data which will determine the magnitude of the answer. Often the wording of the problem will indicate that quantity; for example, "calculate (the answer) given (the specific starting quantity) and - - -", or, "how many - - - are equivalent to (the starting quantity)?" If the problem is worded in a different format and you have difficulty identifying the answer and the specific starting quantity, you may be able to reword the problem into a form like one of the samples above. (These suggestions are designed to help you get started solving problems with the aid of dimensional analysis. They are not a satisfactory substitute for the knowledge of chemical principles and the mathematical skills which you should develop as part of your study of chemistry.) Soon you will single out the answer and the specific starting quantity as part of your reading of the problem without special effort or mechanical aids. When you have completed these first two steps you know the "X" and "Y" of our algebraic example ("X" = the units of the starting quantity and "Y" = the units of the answer.) If you can multiply by the units ratio Y/X you will cancel the units of the starting quantity and supply those of the answer and your problem set-up will be completed. But, you can use a multiplier only when it is a "unity ratio" and if you do not know or have access to the set of numbers which would make the numerator and denominator equivalent you will have to find two (or more) units ratios (which you can express as "unity ratios") whose product will accomplish the same units transform (from starting quantity to answer). If your legs can't move you onto a three foot platform in "one giant stride" you need a set of steps which you can handle! Problem solving by dimensional analysis works the same way. As you grew in physical size you could manage that same three foot rise by a few larger steps and as your problem solving skills develop you will require fewer "unity ratio" steps to solve problems. For example you will use (36 in / 1 yd) instead of (12 in / 1 ft)•(3 ft / 1 yd).

Since our discussion of dimensional analysis has not involved much "chemistry". It is important that you understand that this technique is not a substitute for chemical knowledge and an understanding of chemical principles. I must emphasize that selecting the series of units ratios which will "bridge" between the starting quantity and the answer of a chemistry problem and converting those units ratios into "unity ratios" require chemical knowledge. If you don't know the chemistry, you can't do the problems! But it is important to develop your general problem solving skills using dimensional analysis since you can then utilize this technique to solve problems in any other field after you have gained the necessary knowledge of that discipline.

In this manual exact numbers will be indicated by the use of a star "*", i.e. 12*in/1*ft. The limiting data item with the fewest significant figures (s.f.) will be indicated by a large black dot (•) in a superscript position, i.e. 3.56•in. A large black dot in a mid line position may be used to indicate multiplication i.e. $6.022 \cdot 10^{23}$.

(FOR EXAMPLES TO HELP YOU LEARN TO USE DIMENSIONAL ANALYSIS SEE PAGE 8.)

EXAMPLES

(1.) What is the distance in feet between two points three miles apart?

FIRST - WHAT IS THE ANSWER EXPRESSED IN UNITS? feet

? ft =

SECOND - WHAT IS THE SPECIFIC QUANTITY WHICH WILL DETERMINE THE MAGNITUDE OF THE ANSWER? 3 miles

? ft = 3 mi

WHAT UNITS RATIO MULTIPLIER WOULD SOLVE THE PROBLEM?

$$\frac{\text{feet}}{\text{miles}}$$

$$? \text{ ft} = 3 \text{ mi} \bullet \frac{\text{ft}}{\text{mi}}$$

DO WE KNOW THE NUMBERS WHICH WILL MAKE ft/mi A ''UNITY RATIO''? 5280 feet = 1 mile

$$? \text{ ft} = 3 \text{ mi} \bullet \frac{5280 \text{ ft}}{1 \text{ mi}}$$

CANCEL UNITS
DO ARITHMETIC
DOUBLE UNDERLINE
ANSWER

$$? \text{ ft} = 3 \text{ mi} \bullet \frac{5280 \text{ ft}}{1 \text{ mi}} = \underline{\underline{15{,}840 \text{ ft}}}$$

SINCE WE KNOW THAT 1 mi/ 5280 ft IS ALSO A UNITY RATIO WHAT WOULD HAVE BEEN THE RESULT OF MULTIPLYING BY THAT VALUE? (this question is answered below)

$$? \text{ ft} = 3 \text{ mi} \bullet \frac{1 \text{ mi}}{5280 \text{ ft}} = 0.000568 \frac{\text{mi}^2}{\text{ft}}$$

NOTE THE FOLLOWING The units of the answer do not match the target we set up (? ft). The units do not cancel and we are left with a meaningless MESS! Common sense tells us that the number of feet in the distance of 3 miles must be much larger than 3, not much smaller! (NOTE AT THIS POINT THAT IF WE HAD IGNORED THE UNITS 3 / 5280 LOOKS JUST AS NEAT AS 3 x 5280!)

PLEASE LET THE UNITS HELP YOU SOLVE THE PROBLEM!

IN MANY AREAS OF CHEMISTRY YOUR COMMON SENSE MAY NOT TELL YOU THE MAGNITUDE YOU SHOULD EXPECT FOR THE ANSWER BUT IF YOU TREAT THE UNITS RIGHT THEY WILL ALWAYS KEEP YOU ON THE STRAIGHT AND NARROW PATH TO THE ANSWER.

IN THE PREVIOUS EXAMPLE THE NUMERATOR AND THE DENOMINATOR OF THE UNITY RATIO WERE EQUAL (12 IN = 1 FT). WE ARE NOT LIMITED, HOWEVER, TO THIS CONDITION, BUT CAN ALSO EMPLOY UNITY RATIOS WHOSE NUMERATORS AND DENOMINATORS ARE EQUIVALENT. OFTEN THIS EQUIVALENCY IS ESTABLISHED BY THE SPECIFICATIONS OF THE PROBLEM AND IS VALID ONLY FOR THAT PROBLEM RATHER THAN GENERALLY TRUE LIKE 12 IN = 1 FT. (SEE EXAMPLE BELOW)

(2.) How many hours would be required to cover a distance of 250 miles at a speed of 55 miles per hour. IN THIS PROBLEM WE SHOULD BE ABLE TO IDENTIFY THE UNITS OF THE ANSWER (hours) AND THE SPECIFIC STARTING QUANTITY (250 miles) WITHOUT DIFFICULTY SO WE WILL CONCENTRATE ON THE UNITY RATIO. THE STATED SPEED (55 mph) INFORMS US THAT FOR THIS PROBLEM ONLY A DISTANCE OF 55 MILES IS EQUIVALENT TO A DRIVING TIME OF ONE HOUR (IN THE HYPOTHETICAL CASE THAT THE SPEED COULD BE MAINTAINED AT EXACTLY 55). OUR SET-UP OF THE PROBLEM TO THIS POINT:

$$? \text{ hr} = 250 \text{ mi} \cdot \frac{\text{hr}}{\text{mi}}$$

(THE UNITS SHOW US THAT THIS IS THE RATIO WE WOULD LIKE TO UTILIZE.)

Since FOR THIS PROBLEM ONLY we have an equivalence between 1 hour and 55 miles we can construct a UNITY RATIO of 1 hr/ 55 mi.

$$? \text{ hr} = 250 \text{ mi} \cdot \frac{1 \text{ hr}}{55 \text{ mi}} = \underline{\underline{4.55 \text{ hr}}}$$

(IN THIS SECOND EXAMPLE WE MIGHT BE RELUCTANT TO SAY THAT A DISTANCE AND A TIME ARE EQUAL, BUT FOR THIS PROBLEM, 55 mi and 1 hr ARE EQUIVALENT.

THIS EXAMPLE HAS SHOWN US ANOTHER KEY WORD PER. THE WORD PER WILL OFTEN INDICATE AN EQUIVALENCY WHICH WE CAN USE TO CONSTRUCT UNITY RATIOS.

IN EACH OF THE PREVIOUS EXAMPLES WE WERE ABLE TO SOLVE THE PROBLEM BY THE USE OF A SINGLE UNITY RATIO, BUT MOST OF THE CHEMISTRY PROBLEMS YOU WILL ENCOUNTER WILL REQUIRE MORE THAN ONE TO BRIDGE BETWEEN THE STARTING QUANTITY AND THE ANSWER. (SEE EXAMPLE BELOW)

(3.) How many minutes are required for light to reach the earth from the sun? The distance is about 93,000,000 miles ($9.3 \cdot 10^7$ mi) and the speed of light in a vacuum 186,000 miles per second (1.86

$\bullet\ 10^5$ mi/s). (WE NEED AN ANSWER IN minutes and the starting quantity is the distance ($9.3 \bullet 10^7$ mi). USING THE UNITS OF THESE TWO QUANTITIES INDICATES THE UNITS RATIO WHICH WOULD SOLVE THE PROBLEM IN A SINGLE STEP:

$$? \text{ min} = 9.3 \bullet 10^7 \text{mi} \bullet \frac{\text{min}}{\text{mi}}$$

(WHILE THE PROBLEM INDICATES, by PER AN EQUIVALENCY, IT IS miles and seconds, not miles and minutes. We must play the cards the problem deals to us so we will use the UNITY RATIO available and see how the problem set-up looks at this point:

$$? \text{ min} = 9.3 \bullet 10^7 \text{mi} \bullet \frac{1 \text{ s}}{1.86 \bullet 10^5 \text{mi}} \bullet \frac{\text{min}}{\text{s}}$$

(THE LAST UNITS RATIO IS INDICATED AFTER WE CANCEL UNITS AS THE STEP NEEDED TO COMPLETE THE PROBLEM. Do we know the numbers required to make min/s a UNITY RATIO? Since all of us know that there are 60 seconds in one minute, the answer is YES! The complete problem set-up:

$$? \text{ min} = 9.3^{\bullet} \bullet 10^7 \cancel{\text{mi}} \bullet \frac{1 \cancel{\text{s}}}{1.86 \bullet 10^5 \cancel{\text{mi}}} \bullet \frac{1 \text{ (min)}}{60 \cancel{\text{s}}} = \underline{\underline{8.3 \text{ min}}}$$

THESE THREE EXAMPLES WERE DESIGNED TO GET YOU STARTED AT SOLVING PROBLEMS BY DIMENSIONAL ANALYSIS. YOU MAY WANT TO REFER BACK TO OUR DISCUSSION IF YOU GET "STUCK" ON A CHEMISTRY PROBLEM.

There are some alternative notations for units ratios. The following examples are all equivalent to each other.

$$\frac{\text{g}}{\text{cm}^3} = \text{g/cm}^3 = \text{g cm}^{-3}$$

Please note in the example above (1) $9.3^{\bullet}$ shows by the $^{\bullet}$ that this is the limiting data value (2 s.f.); (2) $\bullet$ is used to indicate multiplication (x is also used in this manual to indicate multiplication); (3) units are canceled with parallel lines to help you find the other unit; (4) the units of the final answer are circled so that you can locate them in the problem; (5) final answers are double underlined (either by ═══ or by ═══). <u>These notations will be used throughout this manual.</u>

PROBLEM SOLUTIONS FOR CHAPTER 1 OF BRADY & HUMISTON 3rd Ed.

1.25 1.0370 (5), 0.000417 (3), 0.00309 (3), 100.1 (4), 9.0010 (5)

1.26 (a) 7.7 (b) 73.3 (c) 0.785 (d) 3.478 (e) 81.4

1.27 (a) 1.25×10^3 (b) 1.3×10^7 (c) 6.023×10^{22} (d) 2.1457×10^5

(e) 3.147×10

1.28 (a) 4.0×10^{-4} (b) 3×10^{-10} (c) 2.146×10^{-3}

(d) 3.28×10^{-5} (e) 9.1×10^{-13}

1.29 (a) 30,000,000,000 (b) 0.0000254 (c) 1.22 (d) 0.00000034

(e) 32,500

1.30 (a) 3.0×10^3m (b) 3.00×10^3m (c) 3.0×10^5cm

1.31 (a) 7.32×10^{14} (b) 1.52×10^4 (c) 1.22×10^{20} (d) 1.7×10^{-4}

(e) 2.70×10^{-25}

1.32 (a) $\underline{5.56 \times 10^{3}}$ (b) $\underline{2.9 \times 10^{4}}$ (c) $\underline{1.49 \times 10^{10}}$ (d) $\underline{3.8 \times 10^{-6}}$ (e) $\underline{9.0 \times 10^{-31}}$

1.33 (a) $\underline{1.638 \times 10^{9}}$ (b) $\underline{2.16 \times 10^{9}}$ (c) $\underline{4.1971 \times 10^{6}}$ (d) $\underline{8.5 \times 10}$ (e) $\underline{1.00}$

1.34 (a) $\underline{2140}$ (b) $\underline{12100}$ (c) $\underline{41}$ (d) $\underline{5.9}$ (e) $\underline{261}$

1.35 (a) $\underline{6.3 \times 10^{-1}}$ (b) $\underline{3.66 \times 10^{4}}$ (c) $\underline{3.03 \times 10^{-2}}$ (d) $\underline{-4.02 \times 10^{-1}}$ (e) $\underline{5.06 \times 10^{-2}}$

1.36 In this problem part (a) asks how many cm equal 1.40 m so the problem is started with "?cm =". This clearly indicates at the start the unit(s) which is our target in the problem. When we encounter that same unit(s) in the set-up of the problem it is circled to remind us that it must not be cancelled. The answer is double underlined.

(THESE METHODS WILL BE USED THROUGHOUT THIS SOLUTIONS MANUAL.)

(a) $$?\text{cm} = 1.40^{\bullet}\,\cancel{\text{m}} \times \frac{10^{2*}\,\text{cm}}{1\,\cancel{\text{m}}} = \underline{\underline{140\ \text{cm}}}$$

(b) $$?\ \text{m} = \overset{(2\ \text{S.F.})}{2800^{\bullet}}\,\cancel{\text{mm}} \times \frac{1^{*}\,\text{m}}{10^{3*}\,\cancel{\text{mm}}} = \underline{\underline{2.8\ \text{m}}}$$

(c) $$?\text{liters} = 185\,\cancel{\text{ml}} \times \frac{1\ \text{liter}}{10^{3}\,\cancel{\text{ml}}} = \underline{\underline{0.185\ \text{liter}}}$$

(1.36 continued) (d) $\underline{1.8 \times 10^{-2}\,\text{kg}}$

(e) $? \text{ m}^2 = 10^{\bullet}\,\text{yd}^2 \times \left(\dfrac{3^*\text{ft}}{1^*\text{yd}}\right)^2 \times \left(\dfrac{1\,\text{m}}{3.281\,\text{ft}}\right)^2 = \underline{8.4\ \text{m}^2}$

(although 10 has only one significant figure by strict interpretation, two were assumed here)

(□ three s.f. assumed)

(f) $?\text{in} = 100^{\square}\,\text{mi} \times \dfrac{5280^*\text{ft}}{1^*\text{mi}} \times \dfrac{12^*\text{in}}{1^*\text{ft}} = \underline{6.34 \times 10^6\ \text{in}}$

(assume 2 s.f.)

(g) $? \text{ mi} = 1^*\text{hr} \times \dfrac{60^*\text{min}}{1^*\text{hr}} \times \dfrac{60^*\text{s}}{1^*\text{min}} \times \dfrac{20^{\bullet}\text{ft}}{1\ \text{s}} \times \dfrac{1^*\text{mi}}{5280^*\text{ft}} = \underline{14\ \text{mi}}$

(h) $\underline{4.1655 \times 10^9\ \text{m}^3}$

(i) $? \text{ cm} = 1^*\text{s} \times \dfrac{1^*\text{min}}{60^*\text{s}} \times \dfrac{1^*\text{hr}}{60^*\text{min}} \times \dfrac{40^{\bullet}\text{mi}}{1^*\text{hr}} \times \dfrac{1609\ \text{m}}{1^*\text{mi}} \times \dfrac{10^{2*}\text{cm}}{1^*\text{m}} = \underline{1.8 \times 10^3\,\text{cm}}$

1.37 (a) nm (b) mm (c) mg (d) cm (e) μl (f) dm^3 or l

1.38 $F = m \bullet a$ $\quad \underline{N = \text{kg} \bullet \text{m} \bullet \text{s}^{-2}}$

1.39 $? \text{ mi} = 1 \text{ gal} \times \frac{3.785 \text{ l}}{1^* \text{gal}} \times \frac{10^\bullet \text{ km}}{1^* \text{l}} \times \frac{1^* \text{mi}}{1.609 \text{ km}} = \underline{24 \text{ mi}}$ (PFERDBERPER)

1.40 $? \text{ km} = 1^* \text{hr} \times \frac{35^\bullet \text{ mi}}{1^* \text{hr}} \times \frac{1.609 \text{ km}}{1^* \text{mi}} = \underline{56 \text{ km}}$

1.41 $? \text{ km} = 4.3^\bullet \text{ mi} \times \frac{1.609 \text{ km}}{1 \text{ mi}} = \underline{6.9 \text{ km}}$

1.42 (th = thrub)
(P) = potato

$? \text{ lb(P)} = 1^* \text{yr} \times \frac{52^\bullet \text{ wk}}{1 \text{ yr}} \times \frac{142^* \text{th}}{1^* \text{wk}} \times \frac{1^* \text{th(P)}}{14 \text{ th}} \times \frac{1^* \text{lb(P)}}{2^* \text{th(P)}} = \underline{260 \text{ lb(P)}}$

1.43 $? \text{ g Mg} = 1^* \text{cm}^3 \text{ Mg} \times \frac{14.3^\bullet \text{ g Mg}}{8.46 \text{ cm}^3 \text{ Mg}} = \underline{1.69 \text{ g Mg}}$ (per cm^3)

DENSITY equals mass per unit volume. The set-up of the problem below starts with: ? g = 1 ml The number of grams is the density in g/ml.

1.44 (M = metal)

$? \text{ g(M)} = 1^* \text{ml (M)} \times \frac{50.8 \text{ g(M)}}{(36.2 - 25.0)^\bullet \text{ ml (M)}} = \underline{4.54 \text{g(M)}}$

(The subtraction of 25.0 from 36.2 yields 11.2, 3 s.f.)

1.45 (volume of a cylinder $= \frac{l \times \pi d^2}{4}$) $?\,cm^3\,Ti = \frac{4.75^{\bullet} cm \times (2.48\ cm)^2}{4}$

$\times\ 3.14159 = \underline{22.94\ cm^3}$ density = $?\ g\,Ti = 1^{*} \cancel{cm^3\,Ti} \times \frac{104.2\ g\,Ti}{22.94\ \cancel{cm^3\,Ti}} =$

$\underline{\underline{4.54\ g\,Ti}}$ (limited to 3 s.f. by 3 s.f. of length, see • above)

(per cm^3)

1.46 (a) $?\ g\,Pb = 12.0^{\bullet} \cancel{cm^3\,Pb} \times \frac{11.35\ g\,Pb}{1^{*} \cancel{cm^3\,Pb}} = \underline{\underline{136\ g\,Pb}}$

(b) $?\ cm^3\,Pb = 155^{\bullet} \cancel{g\,Pb} \times \frac{1^{*} cm^3\,Pb}{11.35\ \cancel{g\,Pb}} = \underline{\underline{13.7\ cm^3\,Pb}}$

1.47 (a) $?\ ml\ CHCl_3 = 10.00^{\bullet} \cancel{g\,CHCl_3} \times \frac{1^{*} ml\ CHCl_3}{1.492\ \cancel{g\,CHCl_3}} = \underline{\underline{6.702\ ml\ CHCl_3}}$

(b) $?\ g\ CHCl_3 = 10.00^{\bullet} \cancel{ml\ CHCl_3} \times \frac{1.492\ g\ CHCl_3}{1^{*} \cancel{ml\ CHCl_3}} = \underline{\underline{14.92\ g\ CHCl_3}}$

(In 1.48a below, the volume of the pycnometer (pycn) is determined by starting with ''What volume equals a pycnometer'', multiplying by the **mass of water** contained in a pycnometer and by the reciprocal of the density of water. In part (b) the mass of the unknown which fills the pycnometer and the volume of the pycnometer (determined in part a) are used as factor-label multipliers.) NOTE - pycn = pycnometer

1.48 (a) $?\ ml = \cancel{1^{*} pycn} \times \frac{(34.914 - 25.296)^{\bullet} \cancel{g\,H_2O}}{\cancel{1^{*} pycn}} \times \frac{1^{*} ml}{0.9970\ \cancel{g\,H_2O}} = \underline{\underline{9.647\ ml}}$

(1.48 continued)

(b) $? \text{ g(un)} = 1^{*}\cancel{\text{ml(un)}} \times \dfrac{(33.485-25.296)\text{g(un)}}{1^{*}\cancel{\text{pycn}}} \times \dfrac{1^{*}\cancel{\text{pycn}}}{9.647^{\bullet}\cancel{\text{ml(un)}}} = \underline{0.8489\text{g(un}}$

1.49 (a) 500 numerical values (b) 105 numerical values

1.50 (IPA = isopropyl alcohol, S.G. = specific gravity)

S.G. of X = density of X / density of water

(a) $\text{S.G. of IPA} = \dfrac{6.56\ \cancel{\text{lb/gal}}\ \text{(IPA)}}{8.34\ \cancel{\text{lb/gal}}\ (H_2O)} = \underline{0.787\ \text{(S.G. of IPA)}}$

(b) $? \text{ g IPA} = 1^{*}\cancel{\text{ml}}\ \text{IPA} \times \dfrac{0.787\ \text{g}/\cancel{\text{ml}}\ \text{(IPA)}}{1.00\ \cancel{\text{g/ml}\ (H_2O)}} \times \dfrac{1.00\ \cancel{\text{g}\ H_2O}}{1^{*}\cancel{\text{ml}\ H_2O}} = \underline{0.787\ \text{g IPA}}$

1.51 (PG = propylene glycol)

$? \text{ lb(PG)} = 10{,}000\ \cancel{\text{gal(PG)}} \times \dfrac{1.04^{\bullet}\text{lb(PG)}}{1.00\ \cancel{\text{lb}\ H_2O}} \times \dfrac{8.34\ \cancel{\text{lb}\ H_2O}}{1^{*}\cancel{\text{gal}}} =$

$\underline{8.67 \bullet 10^{4}\ \text{lb(PG)}}$

1.52 $? \text{ g C} = 24^{\bullet}\cancel{\text{g H}} \times \dfrac{6.0\ \text{g C}}{1.0\ \cancel{\text{g H}}} = \underline{140\ \text{g C}}$

1.53 (I) 10.0•g Cu --- 1.26 g O (II) 10.0•g Cu --- 2.52 g O

1.26 g O(I) / 2.52 g O(II) = 1 (I) / 2 (II)

(to give a more compact set-up at = atom)

1.54
$$?\text{amu} = 1^{*}\,\cancel{\text{C at}} \times \frac{1^{*}\,\text{amu}}{1^{*}\,\cancel{\text{F at}}} \times \frac{12^{*}\,\cancel{\text{F at}}}{19^{*}\,\cancel{\text{C at}}} = \underline{0.6316\ \text{amu}} \left\{\frac{0.0526\ \text{amu}}{1^{*}\text{H at}}\right\}$$

1.55
$$\text{K.E.} = 1/2\ mv^2 \qquad ?\ J = 1/2 \times 4500^{*}\text{kg} \times \left\{\frac{1.79^{\bullet}\text{m}}{1\ \text{s}}\right\}^2 =$$
$$7.21 \bullet 10^3\ \text{kg}\bullet\text{m}^2\bullet\text{s}^{-2} = \underline{7.21\bullet 10^3 \text{J}}$$

$$?\ \text{cal} = 7.21\bullet 10^3\,\cancel{\text{J}} \times \frac{1^{*}\text{cal}}{4.184^{*}\,\cancel{\text{J}}} = \underline{1.72\bullet 10^3}\ \text{cal}$$

1.56
$$?\ J = \frac{145^{\bullet}\,\cancel{\text{lb}}}{2^{*}} \times \left\{\frac{15.3\ \cancel{\text{mi}}}{1^{*}\,\cancel{\text{hr}}}\right\}^2 \times \left\{\frac{1609\ \text{m}}{1^{*}\,\cancel{\text{mi}}}\right\}^2 \times \frac{0.4536\ \text{kg}}{1^{*}\,\cancel{\text{lb}}} \times \left\{\frac{1^{*}\,\cancel{\text{hr}}}{3600\ \text{s}}\right\}^2$$
$$= \underline{1540\ \text{kg}\bullet\text{m}^2\bullet\text{s}^{-2}}\ \underline{(\text{J})}$$

When you work problems involving temperatures it is useful to distinguish between 'temperatures' (i.e. the normal boiling point of water) and temperature changes (i.e. the temperature rise when 1000 calories are absorbed by 25 grams of water) or temperature differences (i.e. the difference between the freezing point of pure water and the freezing point of a 1 molar sugar solution). This can be accomplished by expressing temperatures as °C (°F), 'degrees Celsius' ('degrees Fahrenheit and temperature changes or differences as C°, 'Celsius degrees' (F° 'Fahrenheit degrees'). EXAMPLES: at one atmosphere pressure water boils at 100°C - -

- - - the difference between the freezing point and the boiling point of water is 180 F°. The rules for manipulating these quantities are simple. The difference between two temperatures given in °C (°F) would be expressed in C° (F°). When a temperature change (C° or F°) is added to (subtracted from) a temperature (°C or °F) the result is a new temperature (°C or °F). When temperature changes (differences) are added (subtracted) the result is a new temperature change (difference) so all would be expressed as C° (F°). THIS NOTATION WILL BE USED IN THIS PROBLEM SOLUTIONS MANUAL.

1.57

$$?°F = 30^{\bullet}\,\cancel{°C} \times \frac{9^{*}F°}{5^{*}\cancel{C°}} + 32^{*}F° = \underline{\underline{86°F}}$$

$$?°F = 1983\,\cancel{°C} \times 9^{*}F° / 5^{*}\cancel{C°} + 32^{*}F° = \underline{\underline{3601°F}}$$

1.58

Sample 1

$$?\ gC = 1^{*}\cancel{gF} \times \frac{1.00\ gC}{6.33^{\bullet}\,\cancel{gF}} = \underline{0.158}$$

RATIO C:F

Sample 2

$$?\ gC = 1^{*}gF \times \frac{2.00\ gC}{12.66\ gF} = \underline{0.158}$$

RATIO C:F

Sample 1

$$?\ gC = 1^{*}\cancel{gCl} \times \frac{1.00^{\bullet}\ gC}{11.67\,\cancel{gCl}} = \underline{0.0857}$$

RATIO C:Cl

Sample 2

$$?\ gC = 1^{*}\cancel{gCl} \times \frac{2.00^{\bullet}\ gC}{23.34\,\cancel{gCl}} = \underline{0.0857}$$

RATIO C:Cl

$$?\ gF = 1^{*}\cancel{gCl} \times \frac{6.33^{\bullet}\ gF}{11.67\,\cancel{gCl}} = \underline{0.542}$$

RATIO F:Cl

$$?\ gF = 1^{*}\cancel{gCl} \times \frac{12.66\ gF}{23.34\,\cancel{gCl}} = \underline{0.542}$$

RATIO F:Cl

1.59

Sample 1 | Sample 2 | Sample 3

$$\frac{?\ gX}{1^{*}gY} = \frac{4.31\ gX}{7.69\ gY} = \underline{0.560} \qquad \frac{?\ gX}{1^{*}gY} = \frac{35.9\ gX}{64.1\ gY} = \underline{0.560} \qquad \frac{?\ gX}{1^{*}gY} = \frac{0.718\ gX}{1.282\ gY} = \underline{0.560}$$

1.60 If the formula N_2O has four times as much nitrogen to one part of oxygen as the second formula, then the second formula could be represented as N_2O_4 which would correspond to an empirical formula $\underline{NO_2}$

1.61

$$?\ gP = 1^{*}\cancel{gO} \times \frac{0.845\ gP}{(1.50-0.845)\cancel{gO}} = \underline{1.3} \qquad ?gP = 1^{*}\cancel{gO} \times \frac{1.09\ gP}{(2.50-1/09)\cancel{(g\ O)}}$$

$= \underline{0.773}$ $1.3 / 0.773 = \underline{1.68}$ ∴ RATIO 5:3

1.62

$$?gX = 1^{*}\cancel{at.wt.X} \times \frac{6.92\ g\ X}{0.584\ \cancel{g\ C}} \times \frac{12.0\ \cancel{g\ C}}{1^{*}\cancel{at.wt.C}} \times \frac{1^{*}\cancel{at.wt.C}}{4^{*}\cancel{at.wt.X}} = \underline{35.5\ g\ X}$$

1.63

$$?\ °C = (6152\ \cancel{°F} - 32^{*}\cancel{F°}) \times \frac{5^{*}C°}{9^{*}\cancel{F°}} = \underline{3400.\ °C} \qquad \underline{(K = °C + 273)}$$

∴ $\underline{3673\ K}$

1.64 $?°F = -78\cancel{°C} \times \dfrac{9^{*}F°}{5^{*}\cancel{C°}} + 32F° = \underline{\underline{-108°F}}$ $\quad ?K = -78°C + 273 = \underline{\underline{195\ K}}$

1.65 $?°C = (98.6\cancel{°F} - 32.0\ \cancel{F°}) \times \dfrac{5^{*}C°}{9^{*}\cancel{F°}} = \underline{\underline{37.0°C}}$

$?°F = 39\cancel{°C} \times \dfrac{9^{*}F°}{5^{*}\cancel{C°}} + 32\ F° = \underline{\underline{102°F}}$

(2 s.f.)

1.66 $?\ cal = 500^{*}\cancel{g\ H_2O} \times 24^{\bullet}\cancel{C°} \times \dfrac{1^{*}cal}{1^{*}\cancel{gH_2O} \times 1^{*}\cancel{C°}} = \underline{\underline{12{,}000\ cal}}$

$\left(\underline{\underline{5.0 \times 10^{4}\ J}}\right)$

1.67 $?\ C° = \dfrac{-35.0^{\bullet}\cancel{cal}}{150^{*}\cancel{g\ H_2O}} \times \dfrac{1^{*}C° \times 1^{*}\cancel{g\ H_2O}}{1^{*}\cancel{cal}} = \underline{\underline{-0.233\ C°}}$

$?\ C° = \dfrac{40.0^{\bullet}J}{150^{*}gH_2O} \times \dfrac{1^{*}cal}{4.184^{*}J} \times \dfrac{1^{*}C° \quad \times 1^{*}g\ H_2O}{1^{*}cal} = \underline{\underline{0.0637\ C°}}$

1.68 $?J = \dfrac{255^{\bullet}\cancel{kg}}{2^{*}} \times \left(\dfrac{80.0\cancel{km}}{1^{*}\cancel{hr}}\right)^{2} \times \left(\dfrac{10^{3}{}^{*}\cancel{m}}{1\ \cancel{km}}\right)^{2} \times \left(\dfrac{1^{*}\cancel{hr}}{3600^{*}\cancel{s}}\right)^{2} \times \dfrac{10^{-3}{}^{*}\ kJ}{1^{*}\cancel{kg \bullet m^{2} \bullet s^{-2}}} = \underline{\underline{63.0\ k}}$

$\left(\underline{\underline{6.30 \times 10^{4}\ J}}\right)$

1.69

$$?J = \frac{2.4^{\bullet}\,\text{ton}}{2^*} \times \left\{\frac{35\ \text{mi}}{1^*\text{hr}}\right\}^2 \times \frac{2000^* \times 0.453\ \text{kg}}{1^*\text{ton}} \times \left\{\frac{1609\ \text{m}}{1^*\text{mi}}\right\}^2 \times \left\{\frac{1^*\text{hr}}{3600^*\text{s}}\right\}^2 =$$

$$\underline{\underline{2.7\bullet10^5\ \text{kg}\bullet\text{m}^2\bullet\text{s}^{-2}\quad (\text{J})}}$$

$$?gH_2O = \frac{2.7\bullet10^5\,\text{J}}{10^*\text{C}^\circ} \times \frac{1^*\text{cal}}{4.184^*\text{J}} \times \frac{1^*\text{gH}_2\text{O}\bullet1^*\text{C}^\circ}{1^*\text{cal}} =$$

$$\underline{\underline{6.4\bullet10^3\text{g H}_2\text{O}}}$$

1.70

$$?\ \text{cal} = \frac{1500^*\text{kg}}{2^*}(60^{\bullet}\,\text{m}\bullet\text{s}^{-1})^2 \times \frac{1^*\text{J}}{1^*\text{kg}\bullet\text{m}^2\bullet\text{s}^{-2}} \times \frac{1^*\text{cal}}{4.184^*\text{J}} = \underline{\underline{6.5\bullet10^5\text{cal}}}$$

1.71

$$?\ \text{J} = 510\ \text{N}\bullet40^{\bullet}\,\text{m} \times \frac{1^*\text{kg}\bullet\text{m}\bullet\text{s}^{-2}}{1^*\text{N}} = \underline{\underline{2.0\bullet10^4\ \text{kg}\bullet\text{m}^2\bullet\text{s}^{-2}}}$$

1.72

$$^\circ\text{N} = (\_\_^\circ\text{C} - 80) \times \frac{100\ \text{N}^\circ}{138\ \text{C}^\circ} \quad \text{(FORMULA)}$$

$$?^\circ\text{N} = (0^\circ\text{C} - 80\ \text{C}^\circ) \times \frac{100\ \text{N}^\circ}{138\ \text{C}^\circ} = \underline{\underline{-58^\circ\text{N}}}$$

$$?^\circ\text{N} = (100^\circ\text{C} - 80\ \text{C}^\circ) \times \frac{100\ \text{N}^\circ}{138\ \text{C}^\circ} = \underline{\underline{14^\circ\text{N}}}$$

1.73 HEAT GAINED = HEAT LOST

$$\therefore 3.14\ \cancel{gCu} \times (100\cancel{°C} - X) \times \frac{0.0920\ cal}{1^*\cancel{gCu} \cdot 1^*\cancel{C°}} = 10.00\ \cancel{gH_2O} \times (X - 25.0\cancel{°C}) \times \frac{1^* cal}{1^*\cancel{gH_2O} \cdot 1^*\cancel{C°}}$$

$$(314 - 3.14\ X) \times 0.0920\ cal = (10\ X - 250)\ cal$$

$$278.89 = 10.289\ X \qquad \therefore X = \underline{27.1°C}$$

PROBLEM SOLUTIONS FOR CHAPTER 2 of BRADY & HUMISTON 3rd Ed.

2.16 (a)

$$? \text{ mol S} = 1.00^{\bullet}\,\cancel{\text{mol F}} \times \frac{2\ \text{mol S}}{1\ \cancel{\text{mol F}}} = \underline{\underline{2.00 \text{ mol S}}}$$

(b)

$$? \text{ mol Fe} = 1.44^{\bullet}\,\cancel{\text{mol S}} \times \frac{1\ \text{mol Fe}}{2\ \cancel{\text{mol S}}} = \underline{\underline{0.720 \text{ mol Fe}}}$$

(c)

$$? \text{ mol S} = 3.00^{\bullet}\,\cancel{\text{mol FeS}_2} \times \frac{2\ \text{mol S}}{1\ \cancel{\text{mol FeS}_2}} = \underline{\underline{6.00 \text{ mol S}}}$$

(d)

$$? \text{ mol FeS}_2 = 3.00^{\bullet}\,\cancel{\text{mol Fe}} \times \frac{1\ \text{mol FeS}_2}{1\ \cancel{\text{mol Fe}}} = \underline{\underline{3.00 \text{ mol FeS}_2}}$$

(the word atom is abbreviated 'at')

2.17 (a) $\underline{(SiO_2)_x}$

(b)

$$? \text{ at O} = 25^{*}\,\cancel{\text{at Si}} \times \frac{2\ \text{at O}}{1\ \cancel{\text{at Si}}} = \underline{\underline{50 \text{ at O}}}$$

(c)

$$? \text{ mol O} = 25^{\bullet}\,\cancel{\text{mol Si}} \times \frac{2\ \text{mol O}}{1\ \cancel{\text{mol Si}}} = \underline{\underline{50 \text{ mol O}}}$$

(d)

$$? \text{ mol (Si + O)} = 4.50^{\bullet}\,\cancel{\text{mol SiO}_2} \times \frac{3\ \text{mol (Si + O)}}{1\ \cancel{\text{mol SiO}_2}} = \underline{\underline{13.5 \text{ mol (Si + O)}}}$$

2.18

$$? \text{ mol S} = 1.00^{\bullet}\,\cancel{\text{mol As}_2\text{S}_3} \times \frac{3\ \text{mol S}}{1\ \cancel{\text{mol As}_2\text{S}_3}} = \underline{\underline{3.00 \text{ mol S}}}$$

2.19 $? \text{ mol O} = 1.50 \cancel{\text{mol } Cr_2O_3} \times \dfrac{3 \text{ mol O}}{1 \cancel{\text{mol } Cr_2O_3}} = \underline{4.50 \text{ mol O}}$

2.20 $? \text{ mol } CO_2 = 1 \cancel{\text{mol } CaCO_3} \times \dfrac{1 \text{ mol } CO_2}{1 \cancel{\text{mol } CaCO_3}} = \underline{1 \text{ mol } CO_2}$

2.21 $? \text{ mol } BaSO_4 = 1.25 \cancel{\text{mol } Al_2(SO_4)_3} \times \dfrac{3 \text{ mol } BaSO_4}{1 \cancel{\text{mol } Al_2(SO_4)_3}} = \underline{3.75 \text{ mol}}$ $(BaSO_4)$

2.22 (a) 24.3 gMg (b) 12.0 gC (c) 55.8 gFe (d) 35.4 gCl
(e) 32.0 gS (f) 87.8 gSr

2.23 (a) $? \text{ mol Na} = 50.0 \cancel{\text{g Na}} \times \dfrac{1 \text{ mol Na}}{22.99 \cancel{\text{g Na}}} = \underline{2.17 \text{ mol Na}}$

(b) 0.667 mol As (c) 0.962 mol Cr (d) 1.85 mol Al (e) 1.28 mol K
(f) 0.464 mol Ag

2.24 (a) 40.31 (b) 110.98 (c) 208.22 (d) 135.02 (e) 163.94

2.25 (a) 60.08 (b) 58.33 (c) 246.53 (d) 812.4 (e) 176.13 (f) 342.31

2.26 $? \, g(C) = 1.35^{\bullet} \, \cancel{mol(C)} \times \dfrac{194 \, g(C)}{1^{*} \cancel{mol \, (C)}} = \underline{\underline{262 \, g(C)}}$

2.27 $? g(P) = 2.33^{\bullet} \, \cancel{mol(P)} \times \dfrac{334.22 \, g(P)}{1^{*} \cancel{mol \, (P)}} = \underline{\underline{779 \, g(P)}}$

2.28 $? \, g \, PbSO_4 = 6.30^{\bullet} \, \cancel{mol \, PbSO_4} \times \dfrac{303.2 \, g \, PbSO_4}{1^{*} \cancel{mol \, PbSO_4}} = \underline{\underline{1910 \, g \, PbSO_4}}$ (3 s.f.)

2.29 $? \, g \, TiO_2 = 0.144^{\bullet} \, \cancel{mol \, TiO_2} \times \dfrac{79.90 \, g \, TiO_2}{1^{*} \cancel{mol \, TiO_2}} = \underline{\underline{11.5 \, g \, TiO_2}}$

2.30 $? \, mol \, NaHCO_3 = 242^{\bullet} \, \cancel{g \, NaHCO_3} \times \dfrac{1^{*} mol \, NaHCO_3}{84.0 \, \cancel{g \, NaHCO_3}} = \underline{\underline{2.88 \, mol \, NaHCO_3}}$

2.31 $? \, mol \, (B) = 1.40 \cdot 10^{3\bullet} \, \cancel{g \, (B)} \times \dfrac{1^{*} mol \, (B)}{58.12 \, \cancel{g \, (B)}} = \underline{\underline{24.1 \, mol \, (B)}}$

2.32 $? \, mol \, H_2SO_4 = 85.3^{\bullet} \, \cancel{g \, H_2SO_4} \times \dfrac{1^{*} mol \, H_2SO_4}{98.08 \, \cancel{g \, H_2SO_4}} = \underline{\underline{0.870 \, mol \, H_2SO_4}}$ (3 s.f.)

2.33

$$? \text{ mol PbHAsO}_4 = 25.0^{\bullet} \cancel{\text{g PbHAsO}_4} \times \frac{1^{*} \text{mol PbHAsO}_4}{347.1 \text{ g} \cancel{\text{PbHAsO}_4}} = \underline{\underline{0.0720 \text{ mol (PbHAsO}_4)}}$$

2.34

$$? \text{ mol K} = 125^{\bullet} \cancel{\text{g KCl}} \times \frac{1^{*} \text{mol K}}{74.56 \cancel{\text{g KCl}}} = \underline{\underline{1.68 \text{ mol K}}}$$

2.35

$$? \text{ mol S} = 632^{\bullet} \cancel{\text{gFeS}_2} \times \frac{1^{*} \cancel{\text{mol FeS}_2}}{119.97 \cancel{\text{gFeS}_2}} \times \frac{2^{*} \text{mol S}}{1^{*} \cancel{\text{mol FeS}_2}} = \underline{\underline{10.5 \text{ mol S}}}$$

2.36

$$? \text{mol FeS}_2 = 1.00^{\bullet} \text{kg SO}_2 \times \frac{10^3 {}^{*} \cancel{\text{g SO}_2}}{1^{*} \cancel{\text{kg SO}_2}} \times \frac{1^{*} \cancel{\text{mol SO}_2}}{64.06 \cancel{\text{g SO}_2}} \times \frac{1^{*} \text{mol FeS}_2}{2^{*} \cancel{\text{mol SO}_2}}$$

$$= \underline{\underline{7.81 \text{ mol FeS}_2}}$$

2.37

(a) $$? \text{ g(S)} = 1^{*} \cancel{\text{(M)(S)}} \times \frac{342.30^{\bullet} \text{g (S)}}{6.0225 \cdot 10^{23} \cancel{\text{(M)(S)}}} = \underline{\underline{5.6837 \cdot 10^{-22} \text{g (S)}}}$$

(atom abbreviated 'at' to fit problem on a single line)

(b) $$? \text{ at C} = 1^{*} \cancel{\text{(M)(S)}} \times \frac{5.6837 \cdot 10^{-22} \cancel{\text{g}}}{1^{*} \cancel{\text{(M)(S)}}} \times \frac{6.0225 \cdot 10^{23} \text{at C}}{12.0111 \cancel{\text{g}}} = \underline{\underline{28.499 \text{ at C}}}$$

(Note in part <u>b</u> that a specifier for "g" was omitted in the numerator of one ratio and in the denominator of another. This was not an oversight! Since the relative masses of a sucrose molecule and a carbon atom were requested <u>in this case only</u> grams of two different chemical substances could be canceled.) ((S) = sucrose)

(2.37 part c) $? \text{(M)(S)} = 25.0^{\bullet} \cancel{\text{g(S)}} \times \frac{6.0225 \cdot 10^{23} \text{(M)(S)}}{342.30 \cancel{\text{g (S)}}} = \underline{4.40 \cdot 10^{22} \text{ (M)(S)}}$

((M) = molecule) (at = atom)

(d) $? \text{ at} = 25.0^{\bullet} \cancel{\text{g(S)}} \times \frac{4.40 \cdot 10^{22} \cancel{\text{(M)(S)}}}{25.0 \cancel{\text{g(S)}}} \times \frac{45^{*} \text{at}}{1^{*} \cancel{\text{(M)(S)}}} = \underline{1.98 \cdot 10^{24} \text{ at}}$

(at = atom)

□(There are three moles of carbon atoms per mole of propane.)

2.38 $? \text{at C} = 4.00 \cdot 10^{-8} \cancel{\text{g(P)}} \times \frac{{}^{\square}3 \times 6.022 \cdot 10^{23} \text{at C}}{44.10 \cancel{\text{g (P)}}} = \underline{1.64 \cdot 10^{15} \text{ at C}}$

(at = atom)

2.39 $? \text{gC} = 3.0^{\bullet} \cancel{\text{cmC}} \times \frac{1^{*} \cancel{\text{at C}}}{1.5 \cdot 10^{-8} \cancel{\text{cm C}}} \times \frac{12.01 \text{ g C}}{6.022 \cdot 10^{23} \cancel{\text{at C}}} = \underline{4.0 \cdot 10^{-15} \text{ g C}}$

□(two moles of Cu per mole Cu_2S)

2.40 $? \text{g Cu} = 5.00 \cdot 10^{20} \cancel{S_8\text{(M)}} \times \frac{8^{*} \text{mol} \cancel{Cu_2S}}{6.022 \cdot 10^{23} \cancel{S_8\text{(M)}}} \times \frac{{}^{\square}2 \times 63.54 \text{ g Cu}}{1^{*} \text{mol} \cancel{Cu_2S}} =$

$\underline{0.844 \text{ g Cu}} \leftarrow$ (3 s.f.)

2.41 ? % X = ? g X = 100 g Cpd. (a) $? \text{ gFe} = 100^{*} \cancel{\text{gFeCl}_3} \times \frac{55.85 \text{ gFe}}{162.2 \cancel{\text{gFeCl}_3}}$

= <u>34.43%Fe (65.57% Cl)</u> (b) <u>42.07% Na, 18.89% P, 39.04% O</u>

(c) <u>28.71% K, 0.74% H, 23.55% S, 47.00% O</u> (d) <u>21.21% N, 6.87% H,</u>

(2.41 continued) $\underline{\underline{23.45\%\ P,\ 48.46\%\ O}}$ (e) $\underline{\underline{84.98\%\ Hg,\ 15.02\%\ Cl}}$

2.42

$$\text{(a) ? \% C = ? g C} = 100^{*}\cancel{\text{g Cpd.}} \times \frac{72.06^{\bullet}\,\text{gC}}{78.10\,\cancel{\text{g Cpd.}}} = \underline{\underline{92.27\%\ C,}}$$

$\therefore$ $\underline{\underline{7.73\%\ H}}$ (b) $\underline{\underline{52.14\%\ C,\ 13.13\%\ H,\ 34.73\%\ O}}$ (c) $\underline{\underline{26.58\%\ K,\ 35.35\%\ Cr,\ 38.07\%\ O}}$ (d) $\underline{\underline{63.34\%\ Xe,\ 36.66\%\ F}}$ (e) $\underline{\underline{40.04\%\ Ca,\ 12.00\%\ C,\ 47.96\%\ O}}$

2.43

$$\text{? g N} = 30.0^{\bullet}\cancel{\text{g (G)}} \times \frac{14.01\,\text{g N}}{75.07\,\cancel{\text{g(G)}}} = \underline{\underline{5.60\text{ g N}}}$$

□(three moles H to one mole NH_3)

2.44

$$\text{? g H} = 12.0^{\bullet}\cancel{\text{g }NH_3} \times \frac{{}^{\square}3^{*}\times 1.008\,\text{g H}}{17.03\,\cancel{\text{g }NH_3}} = \underline{\underline{2.13\text{ g H}}}$$

2.45

$$\text{?(SU)} = 1^{*}\cancel{\text{chain}} \times \frac{1^{*}\,\text{(SU)}}{(8\text{x}12 + 8\text{x}1)\cancel{\text{amu}}} \times \frac{10^{6\bullet}\,\cancel{\text{amu}}}{1^{*}\cancel{\text{chain}}} = \underline{\underline{1\bullet 10^{4}\text{ styrene units}}}$$

2.46

$\underline{\underline{S_?O_?}}$

$$\text{? mol S} = 1.40^{\bullet}\cancel{\text{g S}} \times \frac{1^{*}\text{mol S}}{32.06\,\cancel{\text{g S}}} = \underline{0.0437\text{ mol S}}$$

(2.46 continued) $? \text{ mol O} = 2.10^{\bullet} \not{g\ O} \times \frac{1^{*}\text{mol O}}{16.00 \not{g\ O}} = \underline{0.131 \text{ mol O}}$

$S_{0.0437}O_{0.131} = S_{\frac{0.0437}{0.0437}}O_{\frac{0.131}{0.0437}} = \underline{\underline{SO_3}}$

2.47 $\underline{\underline{C_?Cl_?F_?}}$ $?\text{mol C} = 0.423^{\bullet} \not{g\ C} \times \frac{1^{*}\text{mol C}}{12.01 \not{g\ C}} = \underline{0.0352 \text{ mol C}}$

$?\text{mol Cl} = 2.50^{\bullet} \not{g\ Cl} \times \frac{1^{*}\text{mol Cl}}{35.45 \not{g\ Cl}} = \underline{0.0705 \text{ mol Cl}}$ $?\text{mol F} = 1.34^{\bullet} \not{gF}$ $\times \frac{1^{*}\text{mol F}}{19.00 \not{g\ F}} = \underline{0.0705 \text{ mol F}}$ 360:705:705 = 1:2:2 $\therefore$ $\underline{\underline{CCl_2F_2}}$

2.48 $C_?Cl_?$ (assume a 100 g sample of the compound)

$\therefore$ 14.5 g C, 85.5 g Cl $? \text{ mol C} = 14.5^{\bullet} \not{g\ C} \times \frac{1^{*}\text{mol C}}{12.01 \not{g\ C}} = \underline{1.21 \text{ mol C}}$

$? \text{ mol Cl} = 85.5^{\bullet} \not{g\ Cl} \times \frac{1^{*}\text{mol Cl}}{35.45 \not{g\ Cl}} = \underline{2.41 \text{ mol Cl}}$ $C_{1.21}Cl_{2.41} =$

$C_{\frac{1.21}{1.21}}Cl_{\frac{2.41}{1.21}} = \underline{\underline{CCl_2}}$

2.49 $As_?O_?$ $\quad ? \text{ mol As} = 75.7^{\bullet}\cancel{\text{g As}} \times \dfrac{1^{*}\text{mol As}}{74.92\ \cancel{\text{g As}}} = \underline{1.01 \text{ mol As}}$

$$? \text{ mol O} = 24.3^{\bullet}\cancel{\text{g O}} \times \frac{1^{*}\text{mol O}}{16.00\ \cancel{\text{g O}}} = \underline{1.52 \text{ mol O}} \qquad As_{\frac{1.01}{1.01}}O_{\frac{1.52}{1.01}} = As_1O_{1.5} =$$

$\underline{\underline{As_2O_3}}$

2.50 $\quad ? \text{ mol S} = 1.31^{\bullet}\cancel{\text{g S}} \times \dfrac{1^{*}\text{mol S}}{32.06\ \cancel{\text{g S}}} = \underline{0.0409 \text{ mol S}}$

$$? \text{ mol Cl} = (4.22 - 1.31)\ \cancel{\text{g Cl}} \times \frac{1^{*}\text{mol Cl}}{35.45\ \cancel{\text{g Cl}}} = \underline{0.0821 \text{ mol Cl}}$$

$$S_{\frac{0.0409}{0.0401}}Cl_{\frac{0.0821}{0.0401}} = \underline{\underline{SCl_2}}$$

2.51 $Na_?B_?H_?$ $\quad ? \text{ mol Na} = 60.8^{\bullet}\cancel{\text{g Na}} \times \dfrac{1^{*}\text{mol Na}}{22.99\ \cancel{\text{g Na}}} = \underline{2.64 \text{ mol Na}}$

$$? \text{ mol B} = 28.5^{\bullet}\cancel{\text{g B}} \times \frac{1^{*}\text{mol B}}{10.81\ \cancel{\text{g B}}} = \underline{2.64 \text{ mol B}} \qquad ? \text{ mol H} = 10.5^{\bullet}\cancel{\text{g H}} \times$$

$$\frac{1\ \text{mol H}}{1.008\ \cancel{\text{g H}}} = \underline{10.4 \text{ mol H}} \qquad Na_{\frac{2.64}{2.64}}B_{\frac{2.64}{2.64}}H_{\frac{10.4}{2.64}} = Na_1B_1H_{3.95} = \underline{\underline{NaBH_4}}$$

2.52 $C_?H_?O_?$ $\quad ?\ \text{mol C} = 63.2^{\bullet}\,\cancel{\text{g C}} \times \dfrac{1^{*}\,\text{mol C}}{12.01\,\cancel{\text{g C}}} = \underline{5.26\ \text{mol C}}$

$?\ \text{mol H} = 5.26^{\bullet}\,\cancel{\text{g H}} \times \dfrac{1^{*}\,\text{mol H}}{1.008\,\cancel{\text{g H}}} = \underline{5.22\ \text{mol H}}$ $\quad ?\ \text{mol O} = 31.6^{\bullet}\,\cancel{\text{g O}} \times \dfrac{1^{*}\,\text{mol O}}{16.00\,\cancel{\text{g O}}} = \underline{1.98\ \text{mol O}}$

$C_{\frac{5.26}{1.98}}H_{\frac{5.22}{1.98}}O_{\frac{1.98}{1.98}} = C_{2.66}H_{2.64}O_{1.00} = \underline{\underline{C_8H_8O_3}}$ (multiply by 3)

2.53 $?\ \text{mol C} = 1.030^{\bullet}\,\cancel{\text{g CO}_2} \times \dfrac{1^{*}\,\cancel{\text{mol CO}_2}}{44.01\,\cancel{\text{g CO}_2}} \times \dfrac{1^{*}\,\text{mol C}}{1^{*}\,\cancel{\text{mol CO}_2}} = \underline{0.02341\,\cancel{\text{mol C}}}$

$\times \dfrac{12.01^{\bullet}\,\text{g C}}{1^{*}\,\cancel{\text{mol C}}} = \underline{0.2811\ \text{g C}}$ $\quad ?\ \text{mol H} = 0.632^{\bullet}\,\cancel{\text{g H}_2\text{O}} \times \dfrac{1^{*}\,\cancel{\text{mol H}_2\text{O}}}{18.02\,\cancel{\text{g H}_2\text{O}}} \times$

$\dfrac{2^{*}\,\text{mol H}}{1^{*}\,\cancel{\text{mol H}_2\text{O}}} = \underline{0.0702^{\bullet}\,\cancel{\text{mol H}}} \times \dfrac{1.008\ \text{g H}}{1^{*}\,\cancel{\text{mol H}}} = \underline{\underline{0.0708\ \text{g H}}}$ $\quad$ ((S) = sample) $\quad ?\ \text{g O} = 0.537\ \text{g(s)}$

$- 0.2811\ \text{g C} - 0.0708\ \text{g H} = \underline{0.185^{\bullet}\,\cancel{\text{g O}}} \times \dfrac{1^{*}\,\text{mol O}}{16.00\,\cancel{\text{g O}}} = \underline{0.0116\ \text{mol O}}$

$C_{\frac{0.02341}{0.0116}}H_{\frac{0.0702}{0.0116}}O_{\frac{0.0116}{0.0116}} = C_{2.02}H_{6.06}O_{1.00} = \underline{\underline{C_2H_6O}}$

2.54 $?\ \text{mol H} = 0.810^{\bullet}\,\cancel{\text{g H}_2\text{O}} \times \dfrac{2^{*}\,\text{mol H}}{18.02\,\cancel{\text{g H}_2\text{O}}} = \underline{0.0899\ \text{mol H}}$

(2.54 continued)

((S) = sample)

$$? \text{ mol C} = 1.32^{\bullet}\,\cancel{\text{g CO}_2} \times \frac{1^{*}\text{mol C}}{44.01\,\cancel{\text{g CO}_2}} = \underline{0.0300 \text{ mol C}}$$

$$? \text{ mol N} = 1.35\,\cancel{\text{g (S)}} \times \frac{0.284\,\cancel{\text{g NH}_3}}{0.735\,\cancel{\text{g (S)}}} \times \frac{1^{*}\text{mol N}}{17.03\,\cancel{\text{g NH}_3}} = \underline{0.0306 \text{ mol N}}$$

$$? \text{ mol O} = \left(1.35 \text{ g (S)} - 0.0899\,\cancel{\text{mol H}} \times \frac{1.008 \text{ g H}}{1^{*}\cancel{\text{mol H}}} - 0.0300\,\cancel{\text{mol C}} \times \frac{12.01 \text{ g C}}{1^{*}\cancel{\text{mol C}}} - 0.0306\,\cancel{\text{mol N}} \times \frac{14.01 \text{ g N}}{1\,\cancel{\text{mol N}}}\right) \cancel{\text{g O}} \times \frac{1^{*}\text{mol O}}{16.00\,\cancel{\text{g O}}} = \underline{0.0294 \text{ mol O}}$$

$$C_{\frac{0.0300}{0.0294}} H_{\frac{0.0899}{0.0294}} N_{\frac{0.0306}{0.0294}} O_{\frac{0.0294}{0.0294}} = C_{1.02}H_{3.06}N_{1.04}O_{1.00} = \underline{\underline{CH_3NO}}$$

2.55

$$? \text{ g C} = 0.138^{\bullet}\,\cancel{\text{g CO}_2} \times \frac{12.01 \text{ g C}}{44.01\,\cancel{\text{g CO}_2}} = \underline{0.0377 \text{ g C}}$$

$$? \text{ g H} = 0.0566^{\bullet}\,\cancel{\text{g H}_2\text{O}} \times \frac{1^{*}\cancel{\text{mol H}_2\text{O}}}{18.02\,\cancel{\text{g H}_2\text{O}}} \times \frac{2^{*}\cancel{\text{mol H}}}{1^{*}\cancel{\text{mol H}_2\text{O}}} \times \frac{1.008 \text{ g H}}{1^{*}\cancel{\text{mol H}}} = \underline{6.33 \cdot 10^{-3} \text{ g H}}$$

((S) = sample)

$$? \text{ g N} = 0.150^{\bullet}\,\cancel{\text{g(S)}} \times \frac{0.0238\,\cancel{\text{g NH}_3}}{0.200\,\cancel{\text{g (S)}}} \times \frac{14.01 \text{ g N}}{17.03\,\cancel{\text{g NH}_3}} = \underline{0.0147 \text{ g N}}$$

(2.55 continued)

$$? \text{ g Cl} = 0.150^{\bullet}\,\cancel{\text{g (S)}} \times \frac{0.251\ \cancel{\text{g AgCl}}}{0.125\ \cancel{\text{g (S)}}} \times \frac{35.45\ \text{g Cl}}{143.32\ \cancel{\text{g AgCl}}} = \underline{0.0745 \text{ g Cl}}$$

$$? \text{ g C} = 100^{*}\,\cancel{\text{g (S)}} \times \frac{0.0377^{\bullet}\ \text{g C}}{0.150\ \cancel{\text{g (S)}}} = \underline{25.1 \text{ g C}} \quad \underline{\underline{(25.1\% \text{ C})}}$$

$$? \text{ g H} = 100^{*}\,\cancel{\text{g (S)}} \times \frac{6.33^{\bullet} \times 10^{-3}\ \text{g H}}{0.150\ \cancel{\text{g (S)}}} = \underline{4.22 \text{ g H}} \quad \underline{\underline{(4.22\% \text{ H})}}$$

$$? \text{ g N} = 100^{*}\,\cancel{\text{g (S)}} \times \frac{0.0147^{\bullet}\ \text{g N}}{0.150\ \cancel{\text{g (S)}}} = \underline{9.80 \text{ g N}} \quad \underline{\underline{(9.80\% \text{ N})}}$$

$$? \text{ g Cl} = 100^{*}\,\cancel{\text{g (S)}} \times \frac{0.0745^{\bullet}\ \text{g Cl}}{0.150\ \cancel{\text{g (S)}}} = \underline{49.7 \text{ g Cl}} \quad \underline{\underline{(49.7\% \text{ Cl})}}$$

$$? \text{ g O} = 100^{*}\,\cancel{\text{g (S)}} \times \frac{0.0168^{\bullet}\ \text{g O}}{0.150\ \cancel{\text{g (S)}}} = \underline{11.2 \text{ g O}} \quad \underline{\underline{(11.2\% \text{ O})}}$$

(b) $$? \text{ mol C} = 0.0377^{\bullet}\,\cancel{\text{g C}} \times \frac{1^{*}\ \text{mol C}}{12.01\ \cancel{\text{g C}}} = \underline{3.14 \times 10^{-3} \text{mol C}}$$

(continued next page)

(2.55 cont.) $? \text{ mol H} = 6.33^{\bullet} \times 10^{-3} \cancel{\text{g H}} \times \frac{1^{*}\text{mol H}}{1.008 \cancel{\text{g H}}} = \underline{6.28 \times 10^{-3} \text{mol H}}$

$? \text{ mol N} = 0.0147^{\bullet} \cancel{\text{g N}} \times \frac{1^{*}\text{mol N}}{14.01 \cancel{\text{g N}}} = \underline{1.05 \times 10^{-3} \text{mol N}}$ $\quad 0.0745^{\bullet} \cancel{\text{g Cl}} =$

$\underline{2.10 \times 10^{-3} \text{mol Cl}}$, $0.0168^{\bullet} \text{g O} = \underline{1.05 \times 10^{-3} \text{mol O}}$

$C_{\frac{3.14}{1.05}} H_{\frac{6.28}{1.05}} N_{\frac{1.05}{1.05}} O_{\frac{1.05}{1.05}} Cl_{\frac{2.10}{1.05}} = C_{2.99} H_{5.98} N_{1.00} O_{1.00} Cl_{2.00} = \underline{\underline{C_3H_6NOCl_2}}$

2.56 (a) $NaS_2O_3 = \underline{135.1} \therefore \underline{\underline{Na_2S_4O_6}}$ (b) $C_3H_2Cl = \underline{73.5} \therefore$ $\underline{\underline{C_6H_4Cl_2}}$ (c) $C_2HCl = \underline{60.5} \therefore \underline{\underline{C_6H_3Cl_3}}$ (d) $Na_2SiO_3 = \underline{122.1} \therefore$ $\underline{\underline{Na_{12}Si_6O_{18}}}$ (e) $NaPO_3 = \underline{102.0} \therefore \underline{\underline{Na_3P_3O_9}}$

2.57 $? \text{ mol C} = 0.6871^{\bullet} \cancel{\text{g CO}_2} \times \frac{12.01 \text{ g C}}{44.01 \cancel{\text{g CO}_2}} = \underline{0.1875^{\bullet} \cancel{\text{g C}}} \times \frac{1^{*}\text{mol C}}{12.01 \cancel{\text{g C}}}$

(2.57 continued)

$= \underline{0.01561 \text{ mol C}}$ $\quad ?\text{ mol H} = 0.1874^{\bullet}\cancel{\text{g } H_2O} \times \dfrac{1^{*}\cancel{\text{mol } H_2O}}{18.02\cancel{\text{ g } H_2O}} \times \dfrac{2^{*}\text{mol H}}{1\cancel{\text{ mol } H_2O}} =$

$\underline{0.02080^{\bullet}\cancel{\text{mol H}}} \times \dfrac{1.008 \text{ g H}}{1^{*}\cancel{\text{mol H}}} = \underline{0.02097 \text{ g H}}$ $\quad ?\text{ g O} = 0.5000 \text{ g (S)} -$ ((S) = sample)

$0.1875 \text{ g C} - 0.02097 \text{ g H} = \underline{0.2915^{\bullet}\cancel{\text{g O}}} \times \dfrac{1^{*}\text{mol O}}{16.00\cancel{\text{ g O}}} = \underline{0.01822 \text{ mol O}}$

$C_{\frac{0.01561}{0.01561}}H_{\frac{0.02080}{0.01561}}O_{\frac{0.01822}{0.01561}} = C_{1.00}H_{1.33}O_{1.17}$ (multiply by 6) $\therefore$ $\underline{\underline{C_6H_8O_7}}$

2.58 (a) $?\text{ mol } C_2H_2 = 2.50^{\bullet}\cancel{\text{mol } CaC_2} \times \dfrac{1^{*}\text{mol } C_2H_2}{1^{*}\cancel{\text{mol } CaC_2}} = \underline{\underline{2.50 \text{ mol } C_2H_2}}$

(2.58 b) $?\text{ g } C_2H_2 = 0.500^{\bullet}\cancel{\text{mol } CaC_2} \times \dfrac{26.04 \text{ g } C_2H_2}{1^{*}\cancel{\text{mol } CaC_2}} = \underline{\underline{13.0 \text{ g } C_2H_2}}$

(c) $?\text{ mol } H_2O = 3.20^{\bullet}\cancel{\text{mol } C_2H_2} \times \dfrac{2^{*}\text{mol } H_2O}{1^{*}\cancel{\text{mol } C_2H_2}} = \underline{\underline{6.40 \text{ mol } H_2O}}$

(2.58 continued)

(d) $? \text{ g Ca(OH)}_2 = 28.0^{\bullet} \cancel{\text{g C}_2\text{H}_2} \times \frac{74.10 \text{ g Ca(OH)}_2}{26.04 \cancel{\text{g C}_2\text{H}_2}} = \underline{\underline{79.7 \text{ g Ca(OH)}_2}}$

2.59 (a) $? \text{ mol HClO}_3 = 14.3^{\bullet} \cancel{\text{g ClO}_2} \times \frac{1^{*}\cancel{\text{mol ClO}_2}}{67.46 \cancel{\text{g ClO}_2}} \times \frac{5^{*}\text{mol HClO}_3}{6^{*}\cancel{\text{mol ClO}_2}} =$

$\underline{\underline{0.177 \text{ mol HClO}_3}}$ (b) $? \text{ g H}_2\text{O} = 5.74^{\bullet} \cancel{\text{g HCl}} \times \frac{1^{*}\cancel{\text{mol HCl}}}{36.46 \cancel{\text{g HCl}}} \times \frac{3^{*}\cancel{\text{mol H}_2\text{O}}}{1^{*}\cancel{\text{mol HCl}}} \times$

$\frac{18.02 \text{ g H}_2\text{O}}{1^{*}\cancel{\text{mol H}_2\text{O}}} = \underline{\underline{8.50 \text{ g H}_2\text{O}}}$ (c) $? \text{ mol ClO}_2 = 4.25^{\bullet} \cancel{\text{g ClO}_2} \times \frac{1^{*}\text{mol ClO}_2}{67.46 \cancel{\text{g ClO}_2}} =$

$\underline{0.0630 \text{ mol ClO}_2}$ $? \text{ mol H}_2\text{O} = 0.853 \cancel{\text{g H}_2\text{O}} \times \frac{1^{*}\text{mol H}_2\text{O}}{18.02 \cancel{\text{g H}_2\text{O}}} = \underline{0.0473 \text{ mol H}_2\text{O}}$

Since the balanced equation shows that 2 mol of ClO_2 are required for each mole of H_2O (0.0630/0.0473 = 1.33 mol ClO_2 per mol H_2O) ClO_2 is the <u>limiting reactant</u>.

$\therefore ? \text{ g HClO}_3 = 0.0630 \cancel{\text{mol ClO}_2} \times \frac{5^{*}\cancel{\text{mol HClO}_3}}{6^{*}\cancel{\text{mol ClO}_2}} \times \frac{84.46 \text{ g HClO}_3}{1^{*}\cancel{\text{mol HClO}_3}} = \underline{\underline{4.43 \text{ g HClO}_3}}$

2.60 (a) $\underline{P_4 + 5O_2 \rightarrow P_4O_{10}}$

(b) $$?\text{mol } P_4O_{10} = 0.500^{\bullet}\cancel{\text{mol } O_2} \times \frac{1^{*}\text{mol } P_4O_{10}}{5^{*}\cancel{\text{mol } O_2}} = \underline{0.100 \text{ mol } P_4O_{10}}$$

(c) $$?\text{ g } P_4 = 50.0^{\bullet}\cancel{\text{g } P_4O_{10}} \times \frac{123.9 \text{ g } P_4}{283.9\ \cancel{\text{g} P_4O_{10}}} = \underline{21.8 \text{ g } P_4}$$

(d) $$?\text{ g } P_4 = 25.0^{\bullet}\cancel{\text{g } O_2} \times \frac{1^{*}\cancel{\text{mol } O_2}}{32.00\ \cancel{\text{g } O_2}} \times \frac{1^{*}\cancel{\text{mol } P_4}}{5^{*}\cancel{\text{mol } O_2}} \times \frac{123.9 \text{ g } P_4}{1^{*}\cancel{\text{mol } P_4}} = \underline{19.4 \text{ g } P_4}$$

2.61 (a) $$?\text{ mol } HNO_3 = 0.0250^{\bullet}\cancel{\text{mol } N_2H_4} \times \frac{2^{*}\text{mol } HNO_3}{1^{*}\cancel{\text{mol } N_2H_4}} = \underline{0.0500 \text{ mol } HNO_3}$$

(b) $$?\text{ mol } H_2O_2 = 1.35^{\bullet}\cancel{\text{mol } H_2O} \times \frac{7^{*}\text{mol } H_2O_2}{8^{*}\cancel{\text{mol } H_2O}} = \underline{1.18 \text{ mol } H_2O_2}$$

(c) $$?\text{ mol } H_2O = 1.87^{\bullet}\cancel{\text{mol } HNO_3} \times \frac{8^{*}\text{mol } H_2O}{2^{*}\cancel{\text{mol } HNO_3}} = \underline{7.48 \text{ mol } H_2O}$$

(d) $$?\text{ mol } H_2O_2 = 22.0^{\bullet}\cancel{\text{g } N_2H_4} \times \frac{7^{*}\text{mol } H_2O_2}{32.05\ \cancel{\text{g } N_2H_4}} = \underline{4.80 \text{ mol } H_2O_2}$$

(2.61 continued

(e) $? \text{ g } H_2O_2 = 45.8^{\bullet} \cancel{\text{g } HNO_3} \times \frac{1^{*}\cancel{\text{mol } HNO_3}}{63.02 \cancel{\text{ g } HNO_3}} \times \frac{7^{*}\cancel{\text{mol } H_2O_2}}{2^{*}\cancel{\text{mol } HNO_3}} \times \frac{34.02 \text{ g } H_2O_2}{1^{*}\cancel{\text{mol } H_2O_2}} =$ $\underline{86.5 \text{ g } H_2O_2}$

2.62 (a) $? \text{ mol CO} = 35.0^{\bullet} \cancel{\text{mol Fe}} \times \frac{3^{*}\text{mol CO}}{2^{*}\cancel{\text{mol Fe}}} = \underline{52.5 \text{ mol CO}}$

(b) $? \text{ mol } Fe_2O_3 = 4.50^{\bullet} \cancel{\text{mol } CO_2} \times \frac{1^{*}\text{mol } Fe_2O_3}{3^{*}\cancel{\text{mol } CO_2}} = \underline{1.50 \text{ mol } Fe_2O_3}$

(c) $? \text{ g } Fe_2O_3 = 0.570^{\bullet} \cancel{\text{mol Fe}} \times \frac{159.7 \text{ g } Fe_2O_3}{2^{*}\cancel{\text{mol Fe}}} = \underline{45.5 \text{ g } Fe_2O_3}$

(d) $? \text{ mol CO} = 48.5^{\bullet} \cancel{\text{g } Fe_2O_3} \times \frac{3^{*}\text{mol CO}}{159.7 \cancel{\text{ g } Fe_2O_3}} = \underline{0.911 \text{ mol CO}}$

(2.62 e) $? \text{ g Fe} = 18.6^{\bullet} \cancel{\text{g CO}} \times \frac{1^{*}\cancel{\text{mol CO}}}{28.01 \cancel{\text{ g CO}}} \times \frac{2^{*}\cancel{\text{mol Fe}}}{3^{*}\cancel{\text{mol CO}}} \times \frac{55.85 \text{ g Fe}}{1^{*}\cancel{\text{mol Fe}}} =$ $\underline{24.7 \text{ g Fe}}$

2.63 (a) $? \text{ mol } H_2O = 6.50^{\bullet} \cancel{\text{mol } TiCl_4} \times \frac{2^{*}\text{mol } H_2O}{1^{*}\cancel{\text{mol } TiCl_4}} = \underline{13.0 \text{ mol } H_2O}$

(2.63 continued)

(b) $$? \text{ mol HCl} = 8.44^{\bullet}\,\cancel{\text{mol TiCl}_4} \times \frac{4^{*}\,\text{mol HCl}}{1^{*}\,\cancel{\text{mol TiCl}_4}} = \underline{\underline{33.8 \text{ mol HCl}}}$$

(c) $$? \text{ g TiO}_2 = 14.4^{\bullet}\,\cancel{\text{mol TiCl}_4} \times \frac{79.90\ \text{g TiO}_2}{1^{*}\,\cancel{\text{mol TiCl}_4}} = \underline{\underline{1150 \text{ g TiO}_2}}$$

(d) $$? \text{ g HCl} = 85.0^{\bullet}\,\cancel{\text{g TiCl}_4} \times \frac{1^{*}\,\cancel{\text{mol TiCl}_4}}{189.7\,\cancel{\text{g TiCl}_4}} \times \frac{4^{*}\,\cancel{\text{mol HCl}}}{1^{*}\,\cancel{\text{mol TiCl}_4}} \times \frac{36.46\ \text{g HCl}}{1^{*}\,\cancel{\text{mol HCl}}}$$

$\underline{\underline{65.3 \text{ g HCl}}}$

2.64

(assume 4 s.f. intended)

$$? \text{ kg DDT} = 1000^{\bullet}\,\cancel{\text{kg C}_6\text{H}_5\text{Cl}} \times \frac{1^{*}\,\cancel{\text{kg-mol C}_6\text{H}_5\text{Cl}}}{112.57\ \cancel{\text{kg C}_6\text{H}_5\text{Cl}}} \times \frac{354.51\ \text{kg DDT}}{2^{\bullet}\,\cancel{\text{kg-mol C}_6\text{H}_5\text{Cl}}}$$

$= \underline{\underline{1575 \text{ kg DDT}}}$

((A) = aspirin)
((SA) = salicylic acid)

2.65

$$? \text{ g (SA)} = 2 \times 5^{\bullet}\,\cancel{\text{grain (A)}} \times \frac{1^{*}\,\cancel{\text{g}}}{15.4\,\cancel{\text{grains}}} \times \frac{1^{*}\,\cancel{\text{mol (A)}}}{180.0\,\cancel{\text{g (A)}}} \times$$

$$\frac{1^{*}\,\cancel{\text{mol (SA)}}}{1^{*}\,\cancel{\text{mol (A)}}} \times \frac{138\ \text{g (SA)}}{1^{*}\,\cancel{\text{mol (SA)}}} = \underline{\underline{0.5 \text{ g (SA)}}}$$

2.66 $(CH_3)_2NNH_2 + 2N_2O_4 \longrightarrow 4H_2O + 2CO_2 + 3N_2$ (a)

(b) $? \text{ kg } N_2O_4 = 50.0^{\bullet} \text{kg } (CH_3)_2NNH_2 \times \frac{1^{*}\text{kg-mol } (CH_3)_2NNH_2}{60.10 \text{ kg } (CH_3)_2NNH_2} \times$

$\frac{2^{*}\text{kg-mol } N_2O_4}{1^{*}\text{kg-mol } (CH_3)_2NNH_2} \times \frac{92.0 \text{ kg } N_2O_4}{1^{*}\text{kg-mol } N_2O_4} = \underline{150. \text{ kg } N_2O_4}$

((s) = sugar) ((A) = alcohol)

2.67 $? \text{ g (A)} = 500^{\bullet} \text{g (S)} \times \frac{1^{*}\text{mol (S)}}{180.2 \text{ g (S)}} \times \frac{2^{*}\text{mol (A)}}{1^{*}\text{mol (S)}} \times \frac{46.07 \text{ g (A)}}{1^{*}\text{mol (A)}} =$

256 g (A) (maximum)

2.68 (a) $? \text{ mol } C_2H_2 = 35.0^{\bullet} \text{g } C_2H_2 \times \frac{1^{*}\text{mol } C_2H_2}{26.0 \text{ g } C_2H_2} = \underline{1.35 \text{ mol } C_2H_2}$

$? \text{ mol HCl} = 51.0^{\bullet} \text{g HCl} \times \frac{1^{*}\text{mol HCl}}{36.5 \text{ g HCl}} = \underline{1.40 \text{ mol HCl}}$ (since the mol to mol ratio = 1:1, C_2H_2 is limiting)

(b) $? \text{ g } C_2H_3Cl = 1.35^{\bullet} \text{mol } C_2H_2 \times \frac{1^{*}\text{mol } C_2H_3Cl}{1^{*}\text{mol } C_2H_2} \times \frac{62.50 \text{ g } C_2H_3Cl}{1^{*}\text{mol } C_2H_3Cl} = \underline{84.4 \text{g} C_2H_3Cl}$

(2.68 continued)

(c) (0.05 has only 1 s.f.) (answer to 1 s.f.)

$$? \text{ g HCl} = (1.40 - 1.35)^{\bullet}\cancel{\text{mol HCl}} \times \frac{36.5 \text{ g HCl}}{1^{*}\cancel{\text{mol HCl}}} = \underline{\underline{2 \text{ g HCl}}}$$

2.69 (a)

$$? \text{ mol HCl} = 0.430^{\bullet}\cancel{\text{mol COCl}_2} \times \frac{2^{*}\text{mol HCl}}{1^{*}\cancel{\text{mol COCl}_2}} = \underline{\underline{0.860 \text{ mol HCl}}}$$

(b)

$$? \text{ g HCl} = 11.0^{\bullet}\cancel{\text{g CO}_2} \times \frac{1^{*}\cancel{\text{mol CO}_2}}{44.0\cancel{\text{ g CO}_2}} \times \frac{2^{*}\cancel{\text{mol HCl}}}{1^{*}\cancel{\text{mol CO}_2}} \times \frac{36.46 \text{ g HCl}}{1^{*}\cancel{\text{mol HCl}}} = \underline{\underline{18.2 \text{ g HCl}}}$$

(c) Since the mole ratio is 1:1, <u>$COCl_2$ is the limiting reactant.</u>

$$? \text{ mol HCl} = 0.200^{\bullet}\cancel{\text{mol COCl}_2} \times \frac{2^{*}\text{mol HCl}}{1^{*}\cancel{\text{mol COCl}_2}} = \underline{\underline{0.400 \text{ mol HCl}}}$$

2.70 (a)

$$? \text{ g C}_6\text{H}_5\text{Br} = 15.0^{\bullet}\cancel{\text{g C}_6\text{H}_6} \times \frac{157.0 \text{ g C}_6\text{H}_5\text{Br}}{78.11\cancel{\text{ g C}_6\text{H}_6}} = \underline{\underline{30.1 \text{ g C}_6\text{H}_5\text{Br}}}$$

(b)

$$? \text{ g C}_6\text{H}_6 = 2.50^{\bullet}\cancel{\text{g C}_6\text{H}_4\text{Br}_2} \times \frac{78.11 \text{ g C}_6\text{H}_6}{235.9\cancel{\text{ g C}_6\text{H}_4\text{Br}_2}} = \underline{\underline{0.828 \text{ g C}_6\text{H}_6}}$$

<u>(2.70 c)</u> $\underline{\underline{28.5 \text{ g C}_6\text{H}_5\text{Br}}}$

(2.70 continued) (d) The per cent yield equals the number of grams of actual yield per one hundred grams of theoretical yield.

$$?\% = ?\ g(AY) = 100^{*}g\ (TY) \times \frac{28.5^{\bullet}g\ (AY)}{30.1\ g\ (TY)} = \underline{\underline{94.7\%}}$$

2.71 $? \text{ mol } CCl_4 = 150^{\bullet}\not{g\ CCl_4} \times \frac{1^{*}\text{mol } CCl_4}{153.8\ \not{g\ CCl_4}} = \underline{0.975 \text{ mol } CCl_4}$

(assumed 3 s.f.)

$$? \text{ mol } SbF_3 = 100^{\bullet}\not{g\ SbF_3} \times \frac{1^{*}\text{mol } SbF_3}{178.8\ \not{g\ SbF_3}} = \underline{0.559 \text{ mol } SbF_3}$$

Since the mole ratio (CCl_4:SbF_3) is 3:2 and 0.975/0.559 = 1.74, $\underline{SbF_3 \text{ is the limiting reactant.}}$

(a) $? \text{ g } CCl_2F_2 = 0.559^{\bullet}\not{\text{mol } SbF_3} \times \frac{3^{*}\not{\text{mol } CCl_2F_2}}{2^{*}\not{\text{mol } SbF_3}} \times \frac{120.9\ g\ CCl_2F_2}{1^{*}\not{\text{mol } CCl_2F_2}} = \underline{\underline{101 g\ CCl_2F_2}}$

(the difference = 0.137, 3 s.f.)

(b) $? \text{ g } CCl_4 = (0.975 - 1.5\text{x}0.559)\ \not{\text{mol } CCl_4} \times \frac{153.8\ g\ CCl_4}{1^{*}\not{\text{mol } CCl_4}} = \underline{\underline{21.0 \text{ g } CCl_4}}$

(XS reactant)

2.72 $? \text{ mol } C_2H_2 = 5.00^{\bullet}\not{g\ C_2H_2} \times \frac{1^{*}\text{mol } C_2H_2}{26.04\ \not{g\ C_2H_2}} = \underline{\underline{0.192 \text{ mol } C_2H_2}}$

$\underline{1^{st} \text{ STAGE: } 0.192 \text{ mol } Br_2 \text{ (30.7 g) consumed to yield } 0.192 \text{ mol } C_2H_2Br_2}$

(2.72 continued)

(the difference = 9.3, 2 s.f.)

$$? \text{ mol } Br_2 = (40.0 - 30.7)\cancel{\text{g } Br_2} \times \frac{1^*\text{mol } Br_2}{159.8\cancel{\text{g } Br_2}} = \underline{0.058 \text{ mol } Br_2 \text{ (after } 1^{st} \text{ STAGE)}}$$

$$? \text{ g } C_2H_2Br_4 = 0.058^{\bullet}\cancel{\text{mol } Br_2} \times \frac{345.7 \text{ g } C_2H_2Br_4}{1^*\cancel{\text{mol } Br_2}} = \underline{\underline{20 \text{ g } C_2H_2Br_4}} \text{ (}2^{nd}\text{ STAGE)}$$

The net moles of $C_2H_2Br_2$ = moles formed 1st STAGE - moles consumed 2nd.

(difference = 0.134, 3 s.f.)

$$? \text{ g } C_2H_2Br_2 = (0.192 - 0.058)\cancel{\text{mol } C_2H_2Br_2} \times \frac{185.9 \text{ g } C_2H_2Br_2}{1^*\cancel{\text{mol } C_2H_2Br_2}} = \underline{\underline{24.9 \text{ g } C_2H_2Br_2}}$$

2.73 $$? \text{ mol Ag} = 0.950^{\bullet}\cancel{\text{g Ag}} \times \frac{1^*\text{mol Ag}}{107.9\cancel{\text{g Ag}}} = \underline{0.00880 \text{ mol Ag}}$$

$$? \text{ mol } H_2S = 0.140^{\bullet}\cancel{\text{g } H_2S} \times \frac{1^*\text{mol } H_2S}{34.08\cancel{\text{g } H_2S}} = \underline{0.00411 \text{ mol } H_2S}$$

$$? \text{ mol } O_2 = 0.0800^{\bullet}\cancel{\text{g } O_2} \times \frac{1^*\text{mol } O_2}{32.00\cancel{\text{g } O_2}} = \underline{0.00250 \text{ mol } O_2}$$

The mole ratio Ag:H_2S:O_2 = 4:2:1 and 0.00880/0.00411 = 2.14 $\therefore$ Ag is in XS compared to H_2S. Since 0.00411/0.00250 = 1.64, O_2 is in XS compared to H_2S. $\therefore$ $\underline{H_2S \text{ is the limiting reactant.}}$

(2.73 continued)

$$?\text{g } Ag_2S = 0.00411^{\bullet}\cancel{\text{mol } H_2S} \times \frac{2^*\cancel{\text{mol } Ag_2S}}{2^*\cancel{\text{mol } H_2S}} \times \frac{247.8 \text{ g } Ag_2S}{1^*\cancel{\text{mol } Ag_2S}} = \underline{\underline{1.02 \text{ g } Ag_2S}}$$

((WL) = white lead)

2.74

$$? \text{ g (WL)} = 20.0^{\bullet}\cancel{\text{g Pb}} \times \frac{1^*\cancel{\text{mol Pb}}}{207.2\ \cancel{\text{g Pb}}} \times \frac{1^*\cancel{\text{mol (WL)}}}{6^*\cancel{\text{mol Pb}}} \times \frac{775.7 \text{ g (WL)}}{1^*\cancel{\text{mol (WL)}}} =$$

$\underline{\underline{12.5 \text{ g (WL)}}}$ (a)

$$\text{(b) } ? \text{ g } CO_2 = 14.0^{\bullet}\cancel{\text{g } O_2} \times \frac{1^*\cancel{\text{mol } O_2}}{32.00\ \cancel{\text{g } O_2}} \times \frac{2^*\cancel{\text{mol } CO_2}}{3^*\cancel{\text{mol } O_2}} \times \frac{44.01 \text{ g } CO_2}{1^*\cancel{\text{mol } CO_2}} = \underline{\underline{12.8 \text{ g } CO_2}}$$

2.75 (Assume that the mole ratios are 1:1 for all steps.)
((SM) = starting material, (I_1-I_5) = intermediates, (P) = product)

$$?\text{g (P)} = 30.0^{\bullet}\cancel{\text{g (SM)}} \times \frac{1^*\cancel{\text{mol (SM)}}}{80\ \cancel{\text{g (SM)}}} \times \frac{0.5^*\cancel{\text{mol } (I_1)}}{1^*\cancel{\text{mol (SM)}}} \times \frac{0.5^*\cancel{\text{mol } (I_2)}}{1^*\cancel{\text{mol } (I_1)}} \times \frac{0.5^*\cancel{\text{mol } (I_3)}}{1^*\cancel{\text{mol } (I_2)}} \times$$

$$\frac{0.5^*\cancel{\text{mol } (I_4)}}{1^*\cancel{\text{mol } (I_3)}} \times \frac{0.5^*\cancel{\text{mol } (I_5)}}{1^*\cancel{\text{mol } (I_4)}} \times \frac{0.5^*\cancel{\text{mol (P)}}}{1^*\cancel{\text{mol } (I_5)}} \times \frac{100 \text{ g (P)}}{1^*\cancel{\text{mol (P)}}} = \underline{\underline{0.590 \text{g (P)}}}$$

$$? \text{ g (SM)} = 10.0\ \cancel{\text{g (P)}} \times \frac{30^{\bullet} \text{ g (SM)}}{0.59\ \cancel{\text{g (P)}}} = \underline{\underline{510 \text{ g (SM)}}} \quad \text{(2 s.f.)}$$

2.76 $20.8^{\bullet}g\ CH_4 = \underline{1.30\ mol\ CH_4}$ $5.0^{\bullet}g\ CH_3Cl = \underline{0.099\ mol\ CH_3Cl}$

$25.5^{\bullet}g\ CH_2Cl_2 = \underline{0.300\ mol\ CH_2Cl_2}$ $59.0^{\bullet}g\ CHCl_3 = \underline{0.494\ mol\ CHCl_3}$

(the difference, 0.41, has only 2 s.f.)

(a) $? \text{ g } CCl_4 = (1.30-0.099-0.300-0.494)\cancel{\text{mol } CCl_4} \times \frac{153.8 \text{ g } CCl_4}{1^{*}\cancel{\text{mol } CCl_4}} = \underline{\underline{63 \text{ g } CCl_4}}$

(b) $? \text{ g } CCl_4 = 1.30^{\bullet}\cancel{\text{mol } CCl_4} \times \frac{153.8 \text{ g } CCl_4}{1^{*}\cancel{\text{mol } CCl_4}} = \underline{\underline{200. \text{ g } CCl_4}}$ (3 s.f.)

(AY = actual yield, TY = theoretical)

(c) $? \text{ g } CCl_4 \text{ (AY)} = 100^{*}\cancel{\text{g } CCl_4 \text{ (TY)}} \times \frac{63^{\bullet} \text{ g } CCl_4 \text{ (AY)}}{200 \cancel{\text{ g } CCl_4 \text{ (TY)}}} = \underline{\underline{32 \text{ g } CCl_4 \text{ (AY)}}}$ (32%)

(d) $\underline{CH_4 + 4Cl_2 = CCl_4 + 4HCl}$ $? \text{ g } Cl_2 = (0.099+2\bullet0.300+3\bullet0.494+4\bullet0.41)\cancel{\text{mol } Cl_2} \times \frac{70.91 \text{ g } Cl_2}{1^{*}\cancel{\text{mol } Cl_2}} = \underline{\underline{271 \text{ g } Cl_2}}$

(∴ 3 mol Cl_2/mol $CHCl_3$, 2 mol Cl_2/mol CH_2Cl_2, 1 mol Cl_2 per mol CH_3Cl)

2.77 $? \text{ ton } H_2SO_4 = 25.0^{\bullet}\cancel{\text{ton } Ca_3(PO_4)_2} \times \frac{1^{*}\cancel{\text{ton-mol } Ca_3(PO_4)_2}}{310.2 \cancel{\text{ ton } Ca_3(PO_4)_2}} \times \frac{2^{*}\cancel{\text{ton-mol } H_2SO_4}}{1^{*}\cancel{\text{ton-mol } Ca_3(PO_4)_2}} \times \frac{98.08 \text{ ton } H_2SO_4}{1^{*}\cancel{\text{ton-mol } H_2SO_4}} = \underline{\underline{15.8 \text{ ton } H_2SO_4}}$

2.78 (assumed 3 s.f.)

$$? \text{ lb } NH_3 = 650^{\bullet} \cancel{\text{lb } H_2} \times \frac{1^{*} \cancel{\text{lb-mol } H_2}}{2.016 \cancel{\text{lb } H_2}} \times \frac{2^{*} \cancel{\text{lb-mol } NH_3}}{3^{*} \cancel{\text{lb-mol } H_2}} \times \frac{17.03 \text{ (lb } NH_3)}{1^{*} \cancel{\text{lb-mol } NH_3}} =$$

$3660 \text{ lb } NH_3$ (3 s.f.)

2.79 (a) $$? \text{ M NaCl} = \frac{? \text{ mol NaCl}}{1^{*} \text{liter soln.}} = \frac{0.250^{\bullet} \text{ (mol NaCl)}}{0.400 \text{ (liter soln)}} = \underline{0.625 \text{ M NaCl}}$$

((S) = sucrose)

(b) $$? \text{ M (S)} = \frac{1.45^{\bullet} \text{ (mol (S))}}{0.345 \text{ (liter soln)}} = \underline{4.20 \text{ M (S)}}$$

(c) $$? \text{ M } H_2SO_4 = \frac{195^{\bullet} \cancel{\text{g } H_2SO_4}}{0.875 \text{ (liter soln)}} \times \frac{1^{*} \text{ (mol } H_2SO_4)}{98.08 \cancel{\text{g } H_2SO_4}} = \underline{2.27\text{M } H_2SO_4}$$

(d) $$? \text{ M KOH} = \frac{80.0^{\bullet} \cancel{\text{g KOH}}}{0.200 \text{ (liter soln)}} \times \frac{1^{*} \text{ (mol KOH)}}{56.11 \cancel{\text{g KOH}}} = \underline{7.13 \text{ M KOH}}$$

2.80 (a) $$? \text{ M } NH_4Cl = \frac{1.35^{\bullet} \text{ (mol } NH_4Cl)}{2.45 \text{ (liter soln)}} = \underline{0.551 \text{ M } NH_4Cl}$$

(b) $$? \text{ M } AgNO_3 = \frac{0.422^{\bullet} \text{ (mol } AgNO_3)}{0.742 \text{ (liter soln)}} = \underline{0.569 \text{ M } AgNO_3}$$

(2.80 continued)

$$\text{(c) ? M KCl} = \frac{3.00^{\bullet}\times10^{-3}\,\text{mol KCl}}{0.0100\,\text{liter soln}} = \underline{\underline{0.300\text{ M KCl}}}$$

$$\text{(d) ? M NaHCO}_3 = \frac{4.80^{\bullet}\times10^{-2}\,\cancel{\text{g NaHCO}_3}}{0.0250\,\text{liter soln}} \times \frac{1^{*}\text{mol NaHCO}_3}{84.0\,\cancel{\text{g NaHCO}_3}} = \underline{\underline{0.0229\text{ M NaHCO}_3}}$$

2.81

$$\text{(a) ? M NaCl} = \frac{0.250^{\bullet}\,\cancel{\text{mmol}}\,\text{NaCl}}{1^{*}\cancel{\text{ml}}\,\text{soln}} \times \frac{1^{*}\text{mol}}{10^{3*}\,\cancel{\text{mmol}}} \times \frac{10^{3*}\,\cancel{\text{ml}}}{1^{*}\text{liter}} = \frac{0.250\text{ mol NaCl}}{1^{*}\text{liter soln}}$$

(b) ~~0.0250 M~~ 0.2500 M   (c) $\underline{\underline{\text{M = mol/liter = mmol/ml}}}$ (all three are identical)

2.82

$$\text{(a) ?mol Li}_2\text{CO}_3 = 250^{\bullet}\,\cancel{\text{ml soln}} \times \frac{0.250\,\text{mol Li}_2\text{CO}_3}{10^{3}\,\cancel{\text{ml soln}}} = \underline{\underline{0.0625\text{ mol Li}_2\text{CO}_3}}$$

$$\text{(b) ?g Li}_2\text{CO}_3 = 630^{\bullet}\,\cancel{\text{ml soln}} \times \frac{0.250\,\cancel{\text{mol Li}_2\text{CO}_3}}{10^{3*}\,\cancel{\text{ml soln}}} \times \frac{73.89\,\text{g Li}_2\text{CO}_3}{1^{*}\cancel{\text{mol Li}_2\text{CO}_3}} = \underline{\underline{11.6\text{ g Li}_2\text{CO}_3}}$$

$$\text{(c) ? ml soln} = 0.0100^{\bullet}\,\cancel{\text{mol Li}_2\text{CO}_3} \times \frac{10^{3*}\text{ml soln}}{0.250\,\cancel{\text{mol Li}_2\text{CO}_3}} = \underline{\underline{40.0\text{ ml soln}}}$$

$$\text{?ml soln} = 0.0800^{\bullet}\,\cancel{\text{g Li}_2\text{CO}_3} \times \frac{1^{*}\cancel{\text{mmol Li}_2\text{CO}_3}}{0.07389\,\cancel{\text{g Li}_2\text{CO}_3}} \times \frac{1^{*}\text{ml soln}}{0.250\,\cancel{\text{mmol Li}_2\text{CO}_3}} = \underline{\underline{\text{4.33 ml soln}}}$$

2.83 (a) $?\text{ml soln} = 0.100^{\bullet}\cancel{\text{mol KOH}} \times \frac{10^{3*}\text{ml soln}}{0.375\cancel{\text{mol KOH}}} = \underline{\underline{267\text{ ml soln}}}$

(b) $?\text{ mol KOH} = 45.0^{\bullet}\cancel{\text{ml soln}} \times \frac{0.375\text{ mol KOH}}{10^{3*}\cancel{\text{ml soln}}} = \underline{\underline{0.0169\text{ mol KOH}}}$

(c) $?\text{ ml soln} = 10.0^{\bullet}\cancel{\text{g KOH}} \times \frac{1^{*}\cancel{\text{mol KOH}}}{56.11\cancel{\text{g KOH}}} \times \frac{10^{3*}\text{ml soln}}{0.375\cancel{\text{mol KOH}}} = \underline{\underline{475\text{ ml soln}}}$

(d) $?\text{ g KOH} = 1^{*}\cancel{\text{ml soln}} \times \frac{0.375\cancel{\text{mol KOH}}}{10^{3*}\cancel{\text{ml soln}}} \times \frac{56.11\text{ g KOH}}{1^{*}\cancel{\text{mol KOH}}} = \underline{\underline{0.0210\text{ g KOH}}}$

2.84 $?\text{g } Ca(C_2H_3O_2)_2 = 2.00^{\bullet}\cancel{\text{liter soln}} \times \frac{0.250\cancel{\text{mol } Ca(C_2H_3O_2)_2}}{1^{*}\cancel{\text{liter soln}}} \times$

$\frac{158.2\text{ g } Ca(C_2H_3O_2)_2}{1^{*}\cancel{\text{mol } Ca(C_2H_3O_2)_2}} = \underline{\underline{79.1\text{ g } Ca(C_2H_3O_2)_2}}$

2.85 $?\text{g } KNO_3 = 250.0\cancel{\text{ml soln}} \times \frac{3.00^{\bullet} \times 10^{-2}\cancel{\text{mol } KNO_3}}{10^{3*}\cancel{\text{ml soln}}} \times \frac{101.1\text{ g } KNO_3}{1^{*}\cancel{\text{mol } KNO_3}} =$

$\underline{\underline{0.758\text{ g } KNO_3}}$

2.86 $?g\ MgSO_4 \cdot 7H_2O = 500^{\bullet}\ \not{ml\ soln} \times \frac{0.150\ \not{mol\ MgSO_4}}{10^3{}^{*}\ \not{ml\ soln}} \times \frac{246.5\ g\ MgSO_4 \cdot 7H_2O}{1^{*}\ \not{mol\ MgSO_4}}$

$\underline{\underline{18.5\ g\ MgSO_4 \cdot 7H_2O}}$

2.87 $?M\ AgNO_3 = \frac{0.2867^{\bullet}\ \not{g\ AgCl}}{0.02000\ liter\ soln} \times \frac{1^{*}\ mol\ AgNO_3}{143.3\ \not{g\ AgCl}} = \underline{\underline{0.1000\ M\ Ag\ NO_3}}$

3.73 Divide all of the data items by the smallest. This process gives the following set: 1.333, 2.333, 2.666, 1.000, 3.000

The smallest integer multiplier which converts all of the above set to whole numbers is 3. (3.999, 6.999, 7.998, 3.000, 9.000)

Since -2.4×10^{-19}C, the smallest charge represents 3 unit charges the value of the unit charge must be $\underline{-8.0 \times 10^{-20}\ C}$

3.74

^{151}Eu	150.9 amu x 0.4782 = 72.16 amu
^{153}Eu	152.9 amu x 0.5218 = 79.78 amu

AVERAGE ATOMIC MASS OF Eu = 151.94 amu (TOTAL)

3.75

^{10}B	10.01294 amu x 0.196 = 1.96 amu
^{11}B	11.00931 amu x 0.804 = 8.85 amu

AVERAGE ATOMIC MASS OF B = 10.81 amu (TOTAL)

3.76

^{204}Pb	204.0 amu x 0.0148 = 3.02 amu
^{206}Pb	206.0 amu x 0.236 = 48.6 amu
^{207}Pb	207.0 amu x 0.226 = 46.8 amu
^{208}Pb	208.0 amu x 0.523 = 109 amu

AVERAGE ATOMIC MASS OF LEAD = 207 amu (TOTAL)

3.77 let X = fraction ^{35}Cl ∴ 35.453amu =(34.96885X + 36.96590(1-X))

34.96885 X - 36.96590 X = 35.453 - 36.96590 -1.9971 X = -1.513•

X = 0.7576 ∴ 75.76% ^{35}Cl, 24.24% ^{37}Cl

3.78 Let X = fraction ^{107}Ag ∴ (1-X) = fraction ^{109}Ag

106.9041 ~~amu~~ x X + 108.9047 ~~amu~~ x (1-X) = 107.868 ~~amu~~

(106.9041 - 108.9047) X = 107.868 - 108.9047

-2.0006 X = -1.037• ∴ X = 0.5182 ∴ 51.82% ^{107}Ag, 48.18% ^{109}Ag

3.79 $c = \nu\lambda \quad \lambda = c/\nu \quad \nu = c/\lambda$

$$?nm = \frac{3.00 \times 10^{8} \, \cancel{m} \, \cancel{s^{-1}}}{8.0^{\bullet} \times 10^{15} \, \cancel{Hz}} \times \frac{1^{*} \, \cancel{Hz}}{1^{*} \, \cancel{s^{-1}}} \times \frac{10^{9*} \, nm}{1^{*} \, \cancel{m}} = \underline{37 \text{ nm}}$$

$$? \text{ Hz} = \frac{3.00^{\bullet} \times 10^{8} \, \cancel{m} \, \cancel{s^{-1}}}{200.0 \, \cancel{nm}} \times \frac{10^{9*} \, \cancel{nm}}{1^{*} \, \cancel{m}} \times \frac{1^{*} \, Hz}{1^{*} \, \cancel{s^{-1}}} = \underline{1.50 \times 10^{15} \text{ Hz}}$$

3.80 $\lambda = c/\nu \quad \lambda_1 = ? \text{ m} = \dfrac{3.00^{\bullet} \times 10^{8} \, m \, \cancel{s^{-1}}}{101.1 \times 10^{6} \, \cancel{s^{-1}}} = \underline{2.97 \text{ m}}$

$$\lambda_2 = ? \text{ m} = \frac{3.00^{\bullet} \times 10^{8} \, m \, \cancel{s^{-1}}}{880 \times 10^{3} \, \cancel{s^{-1}}} = \underline{3.41 \times 10^{2} \text{ m}}$$

3.81 $\nu = c/\lambda \qquad \nu = ?\ s^{-1} = \frac{3.00^{\bullet} \times 10^{8}\ \cancel{m}\ \textcircled{s^{-1}}}{546 \times 10^{-9}\ \cancel{m}} = \underline{\underline{5.49 \times 10^{14}\ s^{-1}}}$

3.82 $?\ g = 1^{*}\cancel{e^{-}} \times \frac{1^{*}\textcircled{g}}{-1.76^{\bullet} \times 10^{8}\ \cancel{C}} \times \frac{-1.60 \times 10^{-19}\ \cancel{C}}{1^{*}\cancel{e^{-}}} = \underline{\underline{9.09 \times 10^{-28}\ g}}$

3.83 $?p = 1^{*}\cancel{\text{mol H}} \times \frac{1.007\ \cancel{g\ p}}{1^{*}\cancel{\text{mol H}}} \times \frac{9.65^{\bullet} \times 10^{4}\ \cancel{C}}{1^{*}\cancel{g\ p}} \times \frac{1^{*}\textcircled{p}}{1.60 \times 10^{-19}\ \cancel{C}} =$
$\underline{\underline{6.07 \times 10^{23} p}}$

3.84 $\frac{1}{\lambda} = 109{,}678\ cm^{-1}\left(\frac{1}{n_1^2} - \frac{1}{n_2^2}\right)$ 1st line = $6 \rightarrow 5$ = 0.012222
2nd line = $7 \rightarrow 5$ = 0.019592

1st line = $\underline{\underline{7.4599 \times 10^{-4}\ cm}}$ (7459.9 nm)

2nd line = $\underline{\underline{4.6538 \times 10^{-4}\ cm}}$ (4653.8 nm)

3.85 $\frac{1}{\lambda} = 109{,}678\ cm^{-1}\left(\frac{1}{n_2^2} - \frac{1}{n_4^2}\right) = 109{,}678\ cm^{-1}\left(\frac{1}{2^2} - \frac{1}{4^2}\right) =$
$\underline{20{,}564.6\ cm^{-1}}$ $\therefore$ $\lambda = \underline{4.86273 \times 10^{-5}\ cm} = \underline{486.273\ nm}$ $\underline{(6 \rightarrow 3)\ 1094.11 nm}$

3.86 $\Delta E = A\left(\frac{1}{n_1^2} \quad \frac{1}{n_3^2}\right) = A(1 - \frac{1}{9}) = 8/9\ A \qquad \frac{A}{hc} = 109{,}730\ cm^{-1}$

$A = 109{,}730\ \cancel{cm^{-1}} \times 6.626 \times 10^{-34}\ J\ \cancel{s} \times 3.00 \times 10^{8}\ \cancel{m}\ \cancel{s^{-1}} \times \frac{10^{2}\ \cancel{cm}}{1\ \cancel{m}} =$

$\underline{2.181 \times 10^{-18}\ J}$ $\therefore \Delta E = \underline{1.94 \times 10^{-18}\ J}$

3.87 (eq. 3.3) $1/\lambda = 109{,}678\ cm^{-1}\left\{1/n_1^2 - 1/n_2^2\right\}$ $n_1 = 1,\ n_2 = 0$

$E = h\nu \qquad \nu = c/\lambda \qquad \therefore E = hc/\lambda = hc \times 109{,}678\ cm^{-1}\left\{1/n_1^2 - 1/n_2^2\right\}$

$E = 6.626 \times 10^{-34}\ J\ \cancel{s} \times 3.00 \times 10^{10}\ \cancel{cm}\ \cancel{s^{-1}} \times 1.09{,}678\ \cancel{cm^{-1}}\ (1 - 0)$

$E = \underline{2.18 \times 10^{-18} J}$ (MATCHES THE ANSWER TO EXAMPLE 3.6)

3.88 $E = h\nu \quad ?\ J = 6.626 \times 10^{-34}\ J\ \cancel{s} \times 3 \times 10^{15}\ \cancel{Hz} \times \frac{1\ \cancel{s^{-1}}}{1\ \cancel{Hz}} =$

$\underline{2 \times 10^{-18} J}$

3.89 $?\ g\ NUC. = 1\ \cancel{cm^3\ NUC.} \times \frac{1.67 \times 10^{-24}\ g\ NUC.}{4/3\ \pi\ (1.00 \times 10^{-13}/2)^3\ \cancel{cm^3}} =$

$\underline{3.19 \times 10^{15} g\ NUC.}$ (per 1 cm^3)

3.90 VOLUME (sphere) $= 4/3\ \pi\ r^3 = \pi\ d^3/6 \quad \therefore \quad d = (6\ V_{sp}/\pi)^{1/3}$

$V_{sp} = 6.59 \times 10^{21}\ \cancel{ton} \times \frac{2000\ \cancel{lb}}{1\ \cancel{ton}} \times \frac{453.6\ \cancel{g}}{1\ \cancel{lb}} \times \frac{1\ \cancel{cm^3}}{3.19 \times 10^{15}\ \cancel{g}} \times \left(\frac{1\ \cancel{in}}{2.54\ \cancel{cm}}\right)^3 \times$

$\left(\frac{1\ mi}{12 \times 5280\ \cancel{in}}\right)^3 = \underline{4.4822 \times 10^{-4} mi^3} \quad \therefore \quad d_{earth} = \underline{0.0950\ mi}$

3.91 (a) $? \text{ m} = 4.06\ \text{Å} \times \frac{1^* \text{m}}{10^{10}\ \text{Å}} = \underline{4.06 \times 10^{-10} \text{m}}$

(b) $? \text{ nm} = 4.06\ \text{Å} \times \frac{1^* \text{m}}{10^{10}\ \text{Å}} \times \frac{10^{9} \text{nm}}{1^* \text{m}} = \underline{0.406 \text{ nm}}$

(c) $? \text{ pm} = 4.06\ \text{Å} \times \frac{1^* \text{m}}{10^{10}\ \text{Å}} \times \frac{10^{12} \text{pm}}{1^* \text{m}} = \underline{4.06 \times 10^{2} \text{pm}}$

3.92 $? \text{ ton } H_2O = 1.0^{\bullet} \text{g} \times \frac{(3.00^{\bullet} \times 10^{8} \text{m s}^{-1})^2}{100\ C^{\circ}} \times \frac{1^* \text{g}\ H_2O \times 1^* C^{\circ}}{4.184 \text{ J}} \times$

$\frac{1^* \text{kg}}{10^{3*} \text{g}} \times \frac{2.20 \times 10^{-3} \text{ ton}}{2.00 \times 10^{3} \text{ g}} = \underline{2.37 \times 10^{5} \text{ ton } H_2O}$ NOTE: $E = mc^2$

3.93 FOR AN e^-: $\lambda = \frac{h}{m v}$ $\therefore$ $v = \frac{h}{m\lambda}$ $K.E. = \frac{1}{2} mv^2$

$K.E. = \frac{1}{2}m \left(\frac{h}{m \lambda}\right)^2 = \frac{m h^2}{2m^2\lambda^2} = \frac{h^2}{2m\lambda^2} = \frac{(6.626 \times 10^{-34} \text{J s})^2}{2 \times (0.10^{\bullet} \text{nm})^2 \times 9.1091 \times 10^{-31} \text{kg}} \times$

$\left(\frac{10^{9} \text{nm}}{1^* \text{m}}\right)^2 \times \frac{1^* \text{kg m}^2 \text{s}^{-2}}{1^* \text{J}} = \underline{2.4 \times 10^{-17} \text{J}}$

3.94 time = distance / velocity $= \dfrac{d\ m\ \lambda}{h}$ ⟵ $\{v = h/m\lambda\}$ (see 3.93)

$$? \text{ s} = \frac{10^{\bullet}\text{ cm} \times 2.0\text{ g} \times 0.10\text{ nm}}{6.626 \times 10^{-34}\text{ J s}} \times \frac{1^{*}\text{ J}}{1^{*}\text{kg m}^2\text{ s}^{-2}} \times \frac{1^{*}\text{kg}}{10^3\text{ g}} \times \frac{1^{*}\text{m}}{10^2\text{ cm}} \times \frac{1^{*}\text{m}}{10^9\text{ nm}} =$$

3.0×10^{19} s

3.95

$$? \text{ g } H_2O = \frac{1^{*}\text{mol H}}{25^{\bullet}\text{ C}^{\circ}} \times \frac{6.022 \times 10^{23}\ e^-}{1^{*}\text{mol H}} \times \frac{2.180 \times 10^{-18}\text{ J}}{1^{*}e^-} \times$$

$$\frac{1^{*}\text{g } H_2O \times 1^{*}\text{C}^{\circ}}{4.184\text{ J}} = 1.3 \times 10^4 \text{ g } H_2O$$

$\lambda = \dfrac{h}{m \cdot v}$ ← stała Plancka (h); speed (v); masa (m)

4.52

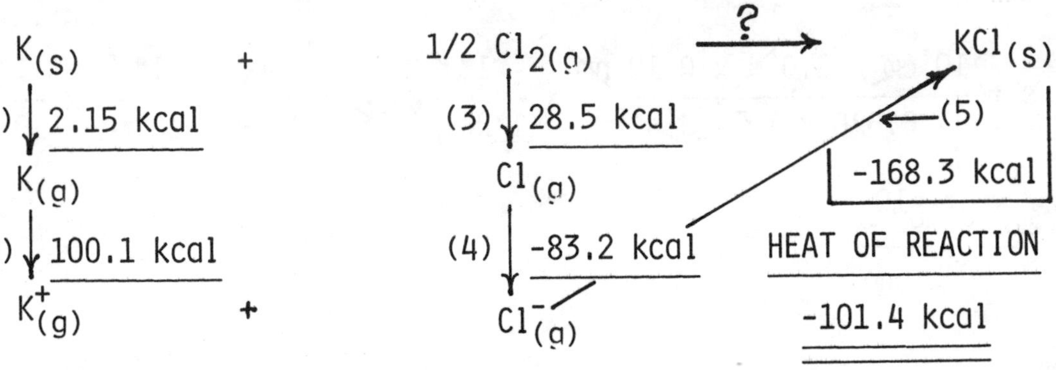

4.53 The difference between an experimental bond energy and a calculated bond energy is proportional to the electronegativity difference between the bonded atoms.

	HF	HCl	HBr	HI
BOND ENERGY DIFFERENCE (kcal/mole)	64	22	12	1

Since the bond energy difference decreases, the electronegativity difference must be decreasing. Since the electronegativity of hydrogen is constant, the electronegativity must be decreasing in the halogen family from fluorine to iodine.

4.54

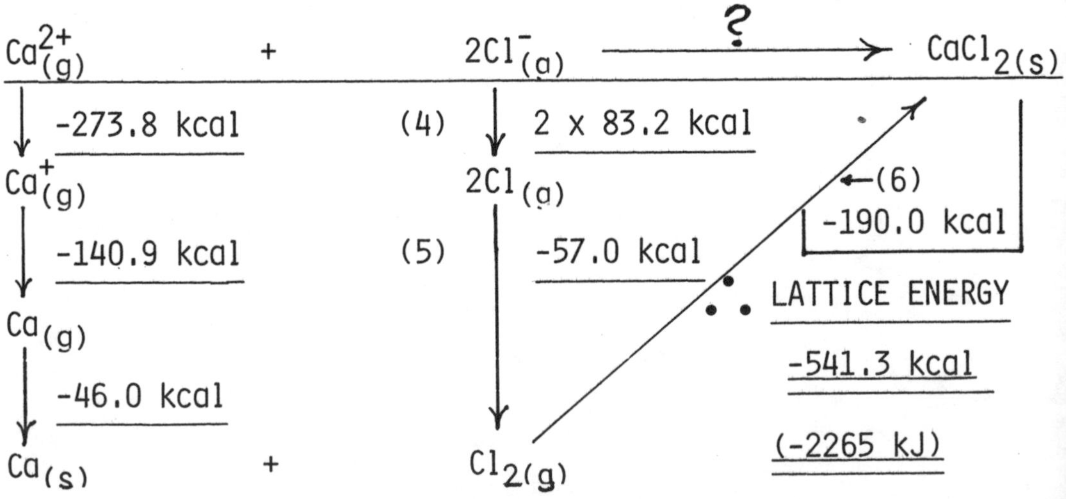

4.55

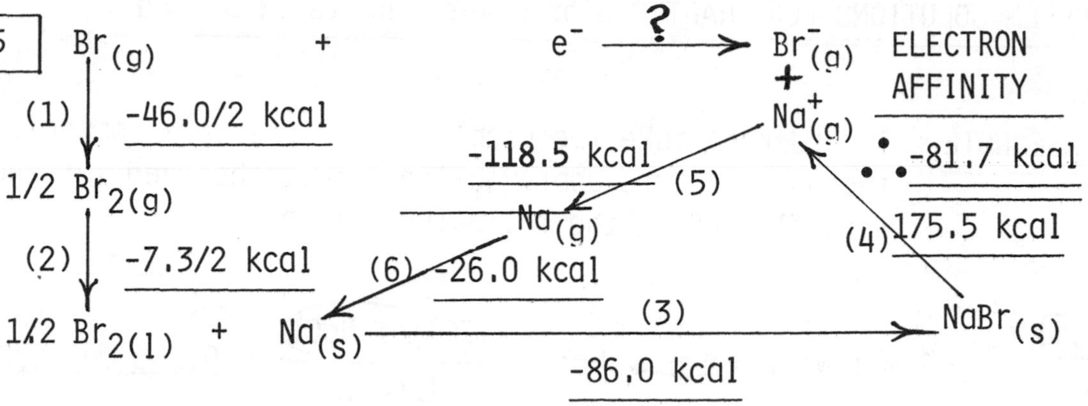

4.56 (refer to problem 4.53)

ALKALI METAL HYDRIDES	LiH	NaH	KH	RbH
BOND ENERGY DIFFERENCE (kcal/mole)	8.2	13	14	19

Since the bond energy difference is increasing, the electronegativity difference must be increasing. Since the electronegativity of hydrogen is constant, and larger than the electronegativity of any alkali metal, the electronegativities of the alkali metals must decrease from lithium to rubidium.

PROBLEM SOLUTIONS FOR CHAPTER 6 of BRADY & HUMISTON 3rd Ed.

PLEASE NOTE: In order to make the problem solutions more compact, the the non-standard abbreviation l. will be used for the word "liter" in the following set-ups.

6.50 (a)

$$? \text{ mol NaCl} = 1^{*}\cancel{\text{l. soln}} \times \frac{1.50^{\bullet}\text{mol NaCl}}{2\ \cancel{\text{l. soln}}} = \underline{0.750 \text{ mol NaCl}} \text{ (M)}$$

(b) 0.992 M (c) 0.556 M (d) 1.36 M (e) 0.962 M

6.51 (a)

$$? \text{ mol KCl} = 250^{\bullet}\cancel{\text{ml soln}} \times \frac{0.100 \text{ mol KCl}}{10^3\ \cancel{\text{ml soln}}} = \underline{2.50 \times 10^{-2} \text{mol KCl}}$$

(b)

$$? \text{ mol } HClO_4 = 1.65^{\bullet}\cancel{\text{l. soln}} \times \frac{1.40 \text{ mol } HClO_4}{1\ \cancel{\text{l. soln}}} = \underline{2.31 \text{ mol } HClO_4}$$

(c)

$$? \text{ mol H } C_2H_3O_2 = 0.0250^{\bullet}\cancel{\text{l. soln}} \times \frac{0.0100 \text{ mol } HC_2H_3O_2}{1\ \cancel{\text{l. soln}}} =$$

$\underline{2.50 \times 10^{-4} \text{ mol } HC_2H_3O_2}$

6.52

$$? \text{ g } Na_2CO_3 = 300^{*}\cancel{\text{ml soln}} \times \frac{0.150^{\bullet}\cancel{\text{mol } Na_2CO_3}}{10^3\ \cancel{\text{ml soln}}} \times \frac{106.0 \text{ g } Na_2CO_3}{1\ \cancel{\text{mol } Na_2CO_3}} =$$

$\underline{4.77 \text{ g } Na_2CO_3}$

6.53 $? \text{ g Ba(OH)}_2 = 250^{*}\cancel{\text{ml soln}} \times \frac{0.300^{*}\cancel{\text{mol OH}^-}}{10^3\ \cancel{\text{ml soln}}} \times \frac{1\ \cancel{\text{mol Ba(OH)}_2}}{2\ \cancel{\text{mol OH}^-}} \times$

$\frac{171\ \text{g Ba(OH)}_2}{1\ \cancel{\text{mol Ba(OH)}_2}} = \underline{\underline{6.41 \text{ g Ba(OH)}_2}}$

6.54 $? \text{ mol HNO}_3 = 1^{*}\cancel{\text{l. soln}} \times \frac{1513^{\bullet}\cancel{\text{g HNO}_3}}{1^{*}\cancel{\text{l. soln}}} \times \frac{1^{*}\text{mol HNO}_3}{63.01\ \cancel{\text{g HNO}_3}} = \underline{\underline{24.01 \text{ mol HNO}_3}}$

(24.01 M)

6.55 Since DENSITY equals grams of solution per milliliter of solution, the problem set-up below begins with "? g soln = 1 ml soln". In this and many other cases, the "=" may be read as "EQUIVALENT TO".

$? \text{ g soln} = 1^{*}\cancel{\text{ml soln}} \times \frac{273.8\ \cancel{\text{g MgSO}_4}}{10^{3*}\cancel{\text{ml soln}}} \times \frac{100^{*}\text{g soln}}{22.0^{\bullet}\cancel{\text{g MgSO}_4}} = \underline{\underline{1.24 \text{ g soln}}}$

∴ The density of the solution is 1.24 g/ml.

The MOLARITY of a solution (M) is the number of moles of solute contained in one liter of solution. The question, "What is the molarity?" asks "How many moles of solute are equivalent to one liter of solution?" The problem set-up below indicates this by; "? M = ? mol = 1 l." at the beginning of the solution.

$? \text{ M} = ? \text{ mol MgSO}_4 = 1^{*}\cancel{\text{l. soln}} \times \frac{273.8\ \cancel{\text{g MgSO}_4}}{1^{*}\cancel{\text{l. soln}}} \times \frac{1^{*}\text{mol MgSO}_4}{120.4\ \cancel{\text{g MgSO}_4}} = \underline{\underline{2.275 \text{ M}}}$

6.56 In this problem, in similar fashion to 6.55, the question What

(6.56 continued)

the percent by weight? asks, How many grams of one component are present in one hundred grams of solution? This is indicated by: ?% = ? g (P) = 100 g soln at the start of the problem set-up. (P)=pollutant, (Bz)=benzene

$$?\% = ?\ g(P) = 100^{*}\cancel{g\ soln} \times \frac{825^{\bullet}\ g\ (P)}{10^{6*}\cancel{g\ soln}} = \underline{0.0825\ g(P)} = \underline{\underline{0.0825\ \%}}$$

$$(b)\ ?\ M = ?\ mol\ (Bz) = 1^{*}\cancel{l.\ soln} \times \frac{10^{3*}\cancel{g\ soln}}{1^{*}\cancel{l.\ soln}} \times \frac{0.0825^{\bullet}\cancel{g\ (Bz)}}{100^{*}\cancel{g\ soln}} \times$$

$$\frac{1^{*}mol\ (Bz)}{78.11\ \cancel{g\ (Bz)}} = \underline{\underline{0.0106\ M}}$$

6.57

$$?\ ml\ NaOH\ soln = 5.00 \times 10^{-3}\ \cancel{mol\ H_2SO_4} \times \frac{2^{*}\cancel{mol\ NaOH}}{1^{*}\cancel{mol\ H_2SO_4}} \times$$

$$\frac{10^{3*}ml\ NaOH\ soln}{0.100^{\bullet}\cancel{mol\ NaOH}} = \underline{\underline{100\ ml\ NaOH\ soln}}$$

6.58

$$?\ ml\ HNO_3\ soln = 3.22^{\bullet}\cancel{g\ Cu} \times \frac{1^{*}\cancel{mol\ Cu}}{63.54\ \cancel{g\ Cu}} \times \frac{8^{*}\cancel{mol\ HNO_3}}{3^{*}\cancel{mol\ Cu}} \times$$

$$\frac{10^{3*}ml\ HNO_3\ soln}{1.250\ \cancel{mol\ HNO_3}} = \underline{\underline{108\ ml\ HNO_3\ soln}}$$

6.59 ? ml $HClO_4$ soln $= 5.25^{\bullet}\,\cancel{g\ Cu(ClO_4)_2} \times \frac{1^{*}\ \cancel{mol\ Cu(ClO_4)_2}}{262.4\ \cancel{g\ Cu(ClO_4)_2}} \times$

$$\frac{2^{*}\ \cancel{mol\ HClO_4}}{1^{*}\ \cancel{mol\ Cu(ClO_4)_2}} \times \frac{10^{3*}\ ml\ HClO_4\ soln}{1.35\ \cancel{mol\ HClO_4}} = \underline{\underline{29.6\ ml\ HClO_4\ soln}}$$

(b) ? g $CuCO_3 = 5.25^{\bullet}\,\cancel{g\ Cu(ClO_4)_2} \times \frac{123.5\ g\ CuCO_3}{262.4\ \cancel{g\ Cu(ClO_4)_2}} = \underline{\underline{2.47\ g\ CuCO_3}}$

6.60 (a) $3NaOH + H_3PO_4 \longrightarrow Na_3PO_4 + 3H_2O$

? ml NaOH soln $= 500^{*}\,\cancel{ml\ H_3PO_4\ soln} \times \frac{0.170^{\bullet}\ \cancel{mol\ H_3PO_4}}{10^{3*}\ \cancel{ml\ H_3PO_4\ soln}} \times \frac{3^{*}\ \cancel{mol\ NaOH}}{1^{*}\ \cancel{mol\ H_3PO_4}} \times$

$$\frac{10^{3*}\ ml\ NaOH\ soln}{0.300\ \cancel{mol\ NaOH}} = \underline{\underline{850\ ml\ NaOH\ soln}}$$

(b) $2NaOH + H_3PO_4 \longrightarrow Na_2HPO_4 + 2H_2O$ $\therefore$ $\underline{\underline{567\ ml\ NaOH\ soln}}$

(c) $NaOH + H_3PO_4 \longrightarrow NaH_2PO_4 + H_2O$ $\therefore$ $\underline{\underline{283\ ml\ NaOH\ soln}}$

6.61 $BaCl_2 + H_2SO_4 \longrightarrow BaSO_4 + 2HCl$? ml $BaCl_2$ soln $= 25.0\ \cancel{ml}$

$$\cancel{H_2SO_4\ soln} \times \frac{0.200\ \cancel{mol\ H_2SO_4}}{10^{3}\ \cancel{ml\ H_2SO_4\ soln}} \times \frac{1^{*}\ \cancel{mol\ BaCl_2}}{1^{*}\ \cancel{mol\ H_2SO_4}} \times \frac{10^{3*}\ ml\ BaCl_2\ soln}{0.100\ \cancel{mol\ BaCl_2}} = \underline{\underline{50.0\ ml\ BaCl_2\ soln}}$$

6.62 $3BaCl_2 + Fe_2(SO_4)_3 \longrightarrow 3BaSO_4 + 2FeCl_3$

$$? \text{ ml } BaCl_2 \text{ soln} = 25.0^{\bullet}\cancel{\text{ml } Fe_2(SO_4)_3 \text{ soln}} \times \frac{0.200\ \cancel{\text{mol } Fe_2(SO_4)_3}}{10^{3*}\cancel{\text{ml } Fe_2(SO_4)_3 \text{ soln}}} \times$$

$$\frac{3^{*}\cancel{\text{mol } BaCl_2}}{1^{*}\cancel{\text{mol } Fe_2(SO_4)_3}} \times \frac{10^{3*}\text{ml } BaCl_2 \text{ soln}}{0.1000\ \cancel{\text{mol } BaCl_2}} = \underline{\underline{150.\ \text{ml } BaCl_2 \text{ soln}}}$$

6.63 Molecular weight is the number of grams per mole. The set-up begins with: "? g = 1 mol". (HBz) = benzoic acid

$$? \text{ g(HBz)} = 1^{*}\cancel{\text{mol(HBz)}} \times \frac{0.244^{\bullet}\text{g (HBz)}}{20.0\ \cancel{\text{ml NaOH soln}}} \times \frac{10^{3*}\cancel{\text{ml NaOH soln}}}{0.100\ \cancel{\text{mol (HBz)}}} = \underline{122\text{g(HBz)}}$$

$\therefore$ $\underline{\underline{MW = 122}}$

6.64 $\underline{\underline{AgNO_3 + NaCl \longrightarrow AgCl_{(s)} + NaNO_3}}$ (a)

(b)

$$?\text{mol } AgNO_3 = 20.0\ \cancel{\text{ml } AgNO_3 \text{ soln}} \times \frac{0.200\ \text{mol } AgNO_3}{10^{3*}\cancel{\text{ml } AgNO_3 \text{ soln}}} = \underline{0.00400 \text{ mol } AgNO_3}$$

LIMITING REACT.

$$? \text{ mol NaCl} = 30.0^{\bullet}\cancel{\text{ml NaCl soln}} \times \frac{0.200\ \text{mol NaCl}}{10^{3*}\cancel{\text{ml NaCl soln}}} = \underline{0.00600 \text{ mol NaCl}}$$

$$\therefore ? \text{ mol AgCl} = 0.00400^{\bullet}\cancel{\text{mol } AgNO_3} \times \frac{1^{*}\text{mol AgCl}}{1^{*}\cancel{\text{mol } AgNO_3}} = \underline{\underline{0.00400 \text{ mol AgCl ppt}}}$$

(6.64 continued)

(c) $? \text{ g AgCl} = 0.00400 \cancel{\text{mol AgCl}} \times \dfrac{143.3 \text{ g AgCl}}{1 \cancel{\text{mol AgCl}}} = \underline{0.573 \text{ g AgCl}}$

(d) For $AgNO_3$: assume that all the Ag^+ (0.004 mol) is consumed and all of the NO_3^- (0.004 mol) remains. For NaCl: assume that all the Na^+ (0.006 mol) remains and that the original Cl^- (0.006 mol) is reduced by (0.004 mol) to a final value of(0.002 mol).

$? \text{ M } (Ag^+) = ? \text{ mol } Ag^+ = 10^3 \text{ ml soln} = \underline{0 \text{ M } (Ag^+)}$

$? \text{ M } (NO_3^-) = ? \text{ mol } (NO_3^-) = 10^3 \cancel{\text{ml soln}} \times \dfrac{0.00400 \text{ mol } NO_3^-}{50.0 \cancel{\text{ml soln}}} = \underline{0.0800 \text{ M}(NO_3^-)}$

$? \text{ M } (Na^+) = ? \text{ mol } Na^+ = 10^3 \cancel{\text{ml soln}} \times \dfrac{0.00600 \text{ mol } Na^+}{50.0 \cancel{\text{ml soln}}} = \underline{0.120 \text{ M } (Na^+)}$

$? \text{ M } (Cl^-) = ? \text{ mol } Cl^- = 10^3 \cancel{\text{ml soln}} \times \dfrac{0.00200 \text{ mol } Cl^-}{50.0 \cancel{\text{ml soln}}} = \underline{0.0400 \text{ M } (Cl^-)}$

6.65 By inspection, HCl is the limiting reactant.

$$HCl + AgNO_3 \longrightarrow AgCl_{(s)} + HNO_3$$

$? \text{ g AgCl} = 25.0 \cancel{\text{ml HCl soln}} \times \dfrac{0.050 \cancel{\text{mol HCl}}}{10^3 \cancel{\text{ml HCl soln}}} \times \dfrac{1 \cancel{\text{mol AgCl}}}{1 \cancel{\text{mol HCl}}} \times \dfrac{143.4 \text{ gAgCl}}{1 \cancel{\text{mol AgCl}}}$

$= \underline{0.18 \text{ g AgCl}}$

6.66 $3BaCl_2 + Fe_2(SO_4)_3 \longrightarrow 3BaSO_{4(s)} + 2FeCl_3$

$$? \text{ mol } BaCl_2 = 50.0^{\bullet} \cancel{\text{ml } BaCl_2 \text{ soln}} \times \frac{0.240 \text{ mol } BaCl_2}{10^3 {}^{*}\cancel{\text{ml } BaCl_2 \text{ soln}}} = \underline{0.0120 \text{ mol } BaCl_2}$$

LIMITING REACTANT

Since 3 mol $BaCl_2$ react with 1 mol $Fe_2(SO_4)_3$ and less than 2:1 are available ∴ $BaCl_2$ is the limiting reactant.

$$? \text{ mol } Fe_2(SO_4)_3 = 45.0^{\bullet} \cancel{\text{ml } Fe_2(SO_4)_3 \text{ soln}} \times \frac{0.180 \text{ mol } Fe_2(SO_4)_3}{10^3 {}^{*}\cancel{\text{ml } Fe_2(SO_4)_3 \text{ soln}}} =$$

$\underline{0.00810 \text{ mol } Fe_2(SO_4)_3}$

(a) $$? \text{ g } BaSO_4 = 0.0120^{\bullet} \cancel{\text{mol } BaCl_2} \times \frac{233.4 \text{ g } BaSO_4}{1^{*}\cancel{\text{mol } BaCl_2}} = \underline{\underline{2.80 \text{ g } BaSO_4}}$$

(b)

ION	MOLES INITIAL	MOLES CONSUMED	MOLES FINAL	MOLAR FINAL CONCENTRATION
Ba^{2+}	0.0120	0.0120	0	0/0.0950 = 0 M
Cl^-	0.0240	0	0.0240	0.0240/0.0950 = 0.253 M
Fe^{3+}	0.0162	0	0.0162	0.0162/0.0950 = 0.171 M
SO_4^{2-}	0.0243	0.0120	0.0123	0.0123/0.0950 = 0.129 M

6.67 $Cr_2(SO_4)_3 + 6NaOH \longrightarrow 2Cr(OH)_3 + 3Na_2SO_4$

$Cr(OH)_3 + 3HNO_3 \longrightarrow Cr(NO_3)_3 + 3H_2O$

(6.67 continued)

$$? \text{ ml } HNO_3 \text{ soln} = 0.500 \cancel{\text{g } Cr_2(SO_4)_3} \times \frac{1 \cancel{\text{mol } Cr_2(SO_4)_3}}{392.2 \cancel{\text{g } Cr_2(SO_4)_3}}$$

$$\times \frac{2 \cancel{\text{mol } Cr(OH)_3}}{1 \cancel{\text{mol } Cr_2(SO_4)_3}} \times \frac{3 \cancel{\text{mol } HNO_3}}{1 \cancel{\text{mol } Cr(OH)_3}} \times \frac{10^3 \text{ ml } HNO_3 \text{ soln}}{0.400 \cancel{\text{mol } HNO_3}} = \underline{\underline{19.1 \text{ ml } HNO_3 \text{ soln}}}$$

6.68 (HCap) = caproic acid

(a) ? MW = ? g (HCap) = 1 ~~mol (HCap)~~

$$\times \frac{0.100 \text{ g (HCap)}}{17.2 \cancel{\text{ml NaOH soln}}} \times \frac{10^3 \cancel{\text{ml NaOH soln}}}{0.0500 \cancel{\text{mol NaOH}}} \times \frac{1 \cancel{\text{mol NaOH}}}{1 \cancel{\text{mol (HCap)}}} = \underline{\underline{116 \text{ g (HCap)} = \text{MW}}}$$

C_3H_6O = 58 ∴ (b) $\underline{\underline{C_6H_{12}O}}$ 2

6.69 XS = excess

$$Ba(OH)_2 + 2HCl \longrightarrow BaCl_2 + 2H_2O$$

$$? \text{ mol HCl} = 500 \cancel{\text{ml HCl}} \times \frac{0.520 \text{ mol HCl}}{10^3 \cancel{\text{ml HCl soln}}} = \underline{0.260 \text{ mol HCl}}$$

$$? \text{ mol } Ba(OH)_2 = 380 \cancel{\text{ml } Ba(OH)_2 \text{ soln}} \times \frac{0.273 \text{ mol } Ba(OH)_2}{10^3 \cancel{\text{ml } Ba(OH)_2 \text{ soln}}} = \underline{0.104 \text{ mol } Ba(OH)_2}$$

ACID TO NEUTRALIZE BASE:

$$? \text{ mol HCl} = 0.104 \cancel{\text{mol } Ba(OH)_2} \times \frac{2 \text{ mol HCl}}{1 \cancel{\text{mol } Ba(OH)_2}} = \underline{0.208 \text{ mol HCl}}$$

? mol HCl XS = 0.260 mol HCl (initial) - 0.208 mol HCl (consumed) =

$\underline{0.052 \text{ mol HCl (final)}}$ (in 0.880 liter) ∴ $\underline{\underline{0.059 \text{ M } H^+}}$

6.70 The determination of an empirical formula requires calculation of the number of moles of each element, and the integer ratio of these numbers. (The mass of Cl can be obtained from the mass of AgCl. The mass of Ti equals the sample mass less the mass of chlorine.)

$Ti_?Cl_?$ $\quad ?\ \text{g Cl} = 0.694^{\bullet}\ \cancel{\text{g AgCl}} \times \dfrac{35.45\ \text{g Cl}}{143.4\ \cancel{\text{g AgCl}}} = \underline{0.172\ \text{g Cl}}$

(s) = SAMPLE

$?\ \text{g Ti} = 0.249\ \text{g (s)} - 0.172\ \text{g Cl} = \underline{0.077\ \text{g Ti}}$

$?\ \text{mol Cl} = 0.172^{\bullet}\ \cancel{\text{g Cl}} \times \dfrac{1^{*}\ \text{mol Cl}}{35.45\ \cancel{\text{g Cl}}} = \underline{0.00484\ \text{mol Cl}}$

$?\ \text{mol Ti} = 0.077^{\bullet}\ \cancel{\text{g Ti}} \times \dfrac{1^{*}\ \text{mol Ti}}{47.9\ \cancel{\text{g Ti}}} = \underline{0.0016\ \text{mol Ti}} \quad Ti_{\frac{0.0016}{0.0016}}Cl_{\frac{0.00484}{0.0016}}$

$Ti_1Cl_{3.0}$ $\therefore$ $\underline{\underline{TiCl_3}}$

6.71 $?\ \text{ppm Hg} = \dfrac{2.1^{\bullet} \times 10^{-5}\ \cancel{\text{mol Hg}}}{25.0\ \cancel{\text{g tuna}}} \times \dfrac{200.6\ \cancel{\text{g Hg}}}{1^{*}\ \cancel{\text{mol Hg}}} \times \dfrac{1^{*}\ \text{ppm Hg}}{1^{*}\ \cancel{\text{g Hg}}(10^{6*}\ \cancel{\text{g tuna}})^{-1}}$

$= \underline{170\ \text{ppm Hg}}$ $\therefore$ $\underline{\underline{\text{CONFISCATED}}}$

6.72 (a) The number of moles of HCl neutralized by the NH_3 are equal to the total moles of HCl less the moles of HCl neutralized by the NaOH.

(6.72 continued)

$$? \text{ mol HCl}_{net} = \left[50.0^{\bullet}\cancel{\text{ml HCl soln}} \times \frac{0.05000 \text{ mol HCl}}{10^3 {}^*\cancel{\text{ml HCl soln}}}\right]$$

$$- \left[30.57\ \cancel{\text{ml NaOH soln}} \times \frac{0.06000 \text{ mol NaOH}}{10^3 {}^*\cancel{\text{ml NaOH soln}}}\right] = \underline{\underline{6.7 \times 10^{-4} \text{ mol HCl}_{net}}}$$

(b) $$? \text{ g N} = 6.7^{\bullet} \times 10^{-4}\ \cancel{\text{mol HCl}} \times \frac{1^*\cancel{\text{mol NH}_3}}{1^*\cancel{\text{mol HCl}}} \times \frac{14.01 \text{ g N}}{1^*\cancel{\text{mol NH}_3}} = \underline{\underline{0.0093 \text{ g N}}}$$

(s) = SAMPLE

(c) $$?\% \text{ N} = ? \text{ g N} = 100^*\text{g (s)} \times \frac{0.0093^{\bullet} \text{ g N}}{0.0500 \text{ g (s)}} = \underline{\underline{19\% \text{ N}}}$$

(gly) = glycine

$$?\% \text{ N(gly)} = ? \text{ g N(gly)} = 100^*\cancel{\text{g (gly)}} \times \frac{14.01^{\bullet} \text{ g N}}{75.08\ \cancel{\text{g(gly)}}} = \underline{18.65\ \% \text{ N(gly)}}$$

rounded to 2 s.f. (19%) the amino acid sample has the same % N as (gly)

6.73 $$3Ba(OH)_2 + Al_2(SO_4)_3 \longrightarrow 3BaSO_{4(s)} + 2Al(OH)_{3(s)}$$

(b) $$? \text{ mol Ba}^{2+} = 40.0^{\bullet}\cancel{\text{ml Ba(OH)}_2 \text{ soln}} \times \frac{0.270 \text{ mol Ba}^{2+}}{10^3 {}^*\cancel{\text{ml Ba(OH)}_2 \text{ soln}}} = \frac{\text{Ba}^{2+}}{0.0108\text{mol}}$$

$\therefore$ $\underline{0.0216 \text{ mol OH}^-}$

$$? \text{ mol SO}_4^{2-} = 25.0^{\bullet}\cancel{\text{ml}Al_2(SO_4)_3 \text{ soln}} \times \frac{0.330 \text{ mol SO}_4^{2-}}{10^3/3\ \cancel{\text{ml Al}_2(SO_4)_3 \text{ soln}}} = \frac{\text{SO}_4^{2-}}{0.0248 \text{ mol}}$$

(6.73 continued) $\therefore$ $\underline{0.0165 \text{ mol } Al^{3+}}$ In view of the mole ratio of the equation, $Ba(OH)_2$ is the limiting reactant.

$$? \text{ g } BaSO_4 = 0.0108^{\bullet}\cancel{\text{mol } Ba^{2+}} \times \frac{233.4 \text{ g } BaSO_4}{1^{*}\cancel{\text{mol } Ba^{2+}}} = \underline{\underline{2.52 \text{ g } BaSO_4}}$$

$$? \text{ g } Al(OH)_3 = 0.0216^{\bullet}\cancel{\text{mol } OH^-} \times \frac{78.00 \text{ g } Al(OH)_3}{3^{*}\cancel{\text{mol } OH^-}} = \underline{\underline{0.562 \text{ g } Al(OH)_3}}$$

$\therefore$ Total wt. of ppt = $\underline{\underline{3.08 \text{ g}}}$ (c) $\underline{\underline{\{Ba^{2+}\} \approx 0 \approx \{OH^-\}}}$ $\underline{\underline{\{Al^{3+}\}}} = (0.0165 - 0.00720) \frac{10^{3*}}{65.0} = \underline{\underline{0.143 \text{ M}}}$ $\underline{\underline{\left[SO_4^{2-}\right]}} = (0.0248 - 0.0108)\frac{10^{3*}}{65.0} = \underline{\underline{0.215 \text{ M}}}$

6.74 Let X = moles of Br^- (X = Moles of AgBr)

$$? \text{ mol } Ag^+ = 1.800^{\bullet}\cancel{\text{g AgCl}} \times \frac{1^{*}\text{mol } Ag^+}{143.3\cancel{\text{ g AgCl}}} = \underline{0.01256 \text{ mol } Ag^+}$$

mol AgCl = mol Ag^+ - X $\therefore$ 2.052 g (ppt) = X•187.78 g AgBr +(0.01256 - X)•143.32 g AgCl 2.052 - 1.800 = 44.46 X $\therefore$ $\underline{X = 0.00567 \text{ (mol } Br^-)}$

$\therefore$ $\underline{0.633 \text{ g } CuBr_2}$

$$?\% \; CuBr_2 = ? \text{ g } CuBr_2 = 100^{*}\cancel{\text{g(s)}} \times \frac{0.633 \text{ g } CuBr_2}{1.850\cancel{\text{ g (s)}}} = \underline{\underline{34.2\% \; CuBr_2}}$$

6.75

$$? \text{ mol Cl}^- = \{0.4881 \text{ g(AgBr, AgI)} - 0.4120 \text{ g(AgBr, AgI, AgCl)}\}\cancel{\text{g increase Cl}\rightarrow\text{Br}} \times \frac{1^*\text{mol Cl}^-}{(79.98 \text{ g Br} - 35.35 \text{ g Cl})\cancel{\text{ g increase Cl}\rightarrow\text{Br}}} = \underline{0.001712 \text{ mol Cl}^-}$$

$$? \text{ g NaCl} = 0.001712\cancel{\text{ mol Cl}^-} \times \frac{58.45 \text{ g NaCl}}{1^*\cancel{\text{mol Cl}^-}} = \underline{0.1001 \text{ g NaCl}}$$

$$? \text{ mol Br}^-\text{(net)} = \{0.586 \text{ g AgI} - 0.4881 \text{ g (AgBr, AgI)}\}\cancel{\text{g increase Br}\rightarrow\text{I}} \times \frac{1^*\text{mol Br}}{(126.9 \text{ g I} - 79.91 \text{ g Br})\cancel{\text{g increase Br}\rightarrow\text{I}}} - 0.001712 \text{ mol (Br}\leftarrow\text{Cl)} = \underline{0.0003884 \text{ mol Br(net)}}$$

$$? \text{ g NaBr} = 0.0003884\cancel{\text{ mol Br}} \times \frac{102.9 \text{ g NaBr}}{1^*\cancel{\text{mol Br}}} = \underline{0.0400 \text{ g NaBr}}$$

$$? \text{ g NaI} = 0.2000 \text{ g(s)} - 0.1001 \text{ g NaCl} - 0.0400 \text{ g NaBr} = \underline{0.0599 \text{ g NaI}}$$

$$?\% \text{ NaCl} = ?\text{g NaCl} = 100^*\cancel{\text{g(s)}} \times \frac{0.1001 \text{ g NaCl}}{0.2000\cancel{\text{ g(s)}}} = \underline{\underline{50.0\% \text{ NaCl}}}$$

$$?\% \text{ NaBr} = ?\text{g NaBr} = 100^*\cancel{\text{g(s)}} \times \frac{0.0400 \text{ g NaBr}}{).2000\cancel{\text{ g(s)}}} = \underline{\underline{20.0\% \text{ NaBr}}} \quad \underline{\underline{(30.0\%\text{NaI})}}$$

6.76 $? \text{ eq } Ba(OH)_2 = 0.200^{\bullet}\cancel{\text{mol } Ba(OH)_2} \times \frac{2^{*}\text{eq } Ba(OH)_2}{1^{*}\cancel{\text{mol } Ba(OH)_2}} = \underline{\underline{0.400 \text{ eq } Ba(OH)_2}}$

6.77 $?\text{mol } H_3PO_4 = 5.00^{\bullet}\cancel{\text{eq } H_3PO_4} \times \frac{1^{*}\text{mol } H_3PO_4}{3^{*}\cancel{\text{eq } H_3PO_4}} = \underline{\underline{1.67 \text{ mol } H_3PO_4}}$

□neutralized to $HAsO_4^{2-}$

6.78 $?\text{eq } H_3AsO_4 = 0.140^{\bullet}\cancel{\text{mol } H_3AsO_4} \times \frac{2^{*}\text{eq } H_3AsO_4^{\square}}{1^{*}\cancel{\text{mol } H_3AsO_4}} = \underline{\underline{0.280 \text{ eq } H_3AsO_4}}$

6.79 (a) $Mn(2+) \longrightarrow Mn(3+) + 1e^-$ $\quad ?g\ MnSO_4 = 1^{*}\cancel{\text{eq } MnSO_4} \times$

$\frac{151^{\bullet}\text{g } MnSO_4}{1^{*}\cancel{\text{eq } MnSO_4}} = \underline{\underline{151 \text{ g } MnSO_4}}$ (b) $(2+) \text{ — } (4+) \therefore$ $\underline{\underline{75.5 \text{ g } MnSO_4}}$

(c) $(2+) \longrightarrow (6+) + 4e^- \therefore$ $\underline{\underline{37.8 \text{ g } MnSO_4}}$ (d) $\underline{\underline{30.2 \text{ g } MnSO_4}}$

6.80 $CrO_4^{2-} + 3e^- \longrightarrow Cr^{3+} \therefore 3 \text{ eq} = 1 \text{ mol}$

$?g\ Na_2CrO_4 = 0.400^{\bullet}\cancel{\text{eq } Na_2CrO_4} \times \frac{162.0 \text{ g } Na_2CrO_4}{3^{*}\cancel{\text{eq } Na_2CrO_4}} = \underline{\underline{21.6 \text{ g } Na_2CrO_4}}$

6.81 $(2+) \longrightarrow (7+) + 5e^- \therefore 5 \text{ eq} = 1 \text{ mol}$

(6.81 continued) $?g\ MnSO_4 \bullet 6H_2O = 300\text{*}\cancel{ml\ soln} \times \frac{0.100\ \cancel{eq\ MnSO_4}}{10^3\text{*}\cancel{ml\ soln}} \times$

$\frac{259.1\ g\ MnSO_4 \bullet 6H_2O}{5\ \cancel{eq\ MnSO_4}} = \underline{\underline{1.55\ g\ MnSO_4 \bullet 6H_2O}}$

6.82 $BiO_3^- \longrightarrow Bi^{3+} - 2e^-$ ∴ $\underline{2\ eq = 1\ mol}$

$Mn^{2+} \longrightarrow MnO_4^- + 5e^-$ ∴ $\underline{5\ eq = 1\ mol}$

$?g\ NaBiO_3 = 0.500\text{*}\cancel{g\ Mn(NO_3)_2} \times \frac{5\text{*}\cancel{eq}}{179\ \cancel{g\ Mn(NO_3)_2}} \times \frac{280\ g\ NaBiO_3}{2\text{*}\cancel{eq}} = \underline{\underline{1.96\ g}}$ $\underline{\underline{NaBiO_3}}$

$^{\Delta}\ H_3PO_4 \rightarrow HPO_4^{2-}$

6.83 (a) $?g\ H_3PO_4 = 1\text{*}\cancel{eq\ H_3PO_4}^{\Delta} \times \frac{98.00\ g\ H_3PO_4}{2\text{*}\cancel{eq}} = \underline{\underline{49.00\ g\ H_3PO_4}}$

(b) $1\text{*}eq\ HClO_4 = 1\text{*}mol\ HClO_4 = \underline{\underline{100.46\ g\ HClO_4}}$ (c) $\underline{IO_3^- \rightarrow I^- - 6e^-}$

$?g\ NaIO_3 = 1\text{*}\cancel{eq\ NaIO_3} \times \frac{197.89\ g\ NaIO_3}{6\text{*}\cancel{eq}} = \underline{\underline{32.98\ g\ NaIO_3}}$

(d) $\underline{IO_3^- \longrightarrow I_2 - 5e^-}$ $?g\ NaIO_3 = 1\text{*}\cancel{eq} \times \frac{197.89\ g\ naIO_3}{5\text{*}\cancel{eq}} = \underline{\underline{39.58\ g}}$ $\underline{\underline{NaIO_3}}$

(e) $3\text{*}eq\ Al(OH)_3 = 1\text{*}mol\ Al(OH)_3$ ∴ $1\text{*}eq\ Al(OH)_3 = \underline{\underline{26.00\ g\ Al(OH)_3}}$

6.84

(a) $? \text{ N} = ? \text{ eq} = 1^* \cancel{\text{liter soln}} \times \frac{22.0^{\bullet} \cancel{\text{g Sr(OH)}_2}}{0.800 \cancel{\text{liter soln}}} \times \frac{2^* \text{eq Sr(OH)}_2}{121.6 \cancel{\text{g Sr(OH)}_2}}$

$= \underline{\underline{0.452 \text{ N}}}$

(b) $? \text{ N} = ? \text{ eq} = 1^* \text{liter soln} \times \frac{0.25^{\bullet} \cancel{\text{mol } H_2SO_4}}{1^* \text{liter soln}} \times \frac{2^* \text{eq } H_2SO_4}{1^* \cancel{\text{mol } H_2SO_4}}$

$= \underline{\underline{0.50 \text{ N}}}$

$^{\Delta} H_3PO_4 \longrightarrow HPO_4^{2-}$

(c) $? \text{ N} = ? \text{ eq} = 1^* \cancel{\text{liter soln}} \times \frac{0.150^{\bullet} \cancel{\text{mol } H_3PO_4}}{1^* \cancel{\text{liter soln}}} \times \frac{2^* \text{eq } H_3PO_4^{\Delta}}{1^* \cancel{\text{mol } H_3PO_4}} = \underline{\underline{0.300 \text{ N}}}$

$Cr_2O_7^{2-} \rightarrow 2Cr^{3+} + 6e^-$

(d) $? \text{ N} = ? \text{ eq} = 1^* \cancel{\text{liter soln}} \times \frac{41.7^{\bullet} \cancel{\text{g } K_2Cr_2O_7}}{0.600 \cancel{\text{liter soln}}} \times \frac{6^* \text{eq } K_2Cr_2O_7}{294.2 \cancel{\text{g } K_2Cr_2O_7}} =$

$\underline{\underline{1.42 \text{ N}}}$

(e) $Na_2O + H_2O \longrightarrow 2 \text{ NaOH}$ $\quad ?\text{N} = ?\text{eq} = 1^* \cancel{\text{liter soln}} \times \frac{25.0^{\bullet} \cancel{\text{g } Na_2O}}{1.50 \cancel{\text{liter}}} \times$

$\frac{2^* \text{eq NaOH}}{61.98 \cancel{\text{g } Na_2O}} = \underline{\underline{0.538 \text{ N}}}$

(f) $?\text{N} = ?\text{eq} = 1^* \cancel{\text{liter soln}} \times \frac{0.135 \text{ eq } H_2SO_4}{0.400 \cancel{\text{liter soln}}}$

$= \underline{\underline{0.338 \text{ N}}}$

6.85 The equivalent weight of the acid is the number of grams per equivalent. Equivalents of acid equal equivalents of base. The number of equivalents of base can be determined from the number of milliliters of solution and the normality of the solution.

ACID =(A) BASE = (B)

$? \text{ g(A)} = 1^* \cancel{\text{eq(A)}} \times \frac{4.93 \text{ g(A)}}{0.129 \cancel{\text{liter (B)}}} \times \frac{1^* \cancel{\text{liter (B)}}}{0.850 \cancel{\text{eq(B)}}} \times \frac{1^* \cancel{\text{eq(B)}}}{1^* \cancel{\text{eq(A)}}} = \underline{\underline{45.0 \text{ g(A)}}}$

6.86 This problem requires calculation of the number of milliliters of solution which will oxidize a measured volume of the other solution. First the given data is used to determine the molarity of the $KMnO_4$ solution. $C_2O_4^{2-} \longrightarrow 2CO_2 + 2e^-$ (2 eq = 1 mol)

$MnO_4^- + 5e^- \longrightarrow Mn^{2+}$ (5 eq = 1 mol)

$$?M = ?\ \text{mol } KMnO_4 = 1^*\text{liter soln(1)} \times \frac{0.0500^\bullet\ \text{liter soln(2)}}{0.0450\ \text{liter soln(1)}} \times \frac{0.250\ \text{eq } C_2O_4^{2-}}{1^*\text{liter soln(2)}}$$

soln(1) = $KMnO_4$ soln(2) = $H_2C_2O_4$

$$\times \frac{1^*\text{eq } MnO_4^-}{1^*\text{eq } C_2O_4^{2-}} \times \frac{1^*\text{mol } KMnO_4}{5^*\text{eq } MnO_4^-} = \underline{0.0556\ M}\ (KMnO_4)$$

Second, the volume of solution(1) (which has a different effective normality in that reaction) which will oxidize the specified volume of solution(3) is calculated. $MnO_4^- + 3e^- \longrightarrow MnO_2$ (3 eq = 1 mol)

$$?\ \text{ml soln(1)} = 25.0^\bullet\ \text{ml soln(3)} \times \frac{0.250\ \text{eq } C_2O_4^{2-}}{10^{3*}\text{ml soln(3)}} \times \frac{1^*\text{eq } MnO_4^-}{1^*\text{eq } C_2O_4^{2-}} \times \frac{1^*\text{mol } MnO_4^-}{3^*\text{eq } MnO_4^-}$$

$$\times \frac{10^{3*}\text{ml soln(1)}}{0.0556\ \text{mol } MnO_4^-} = \underline{\underline{37.5\ \text{ml soln(1)}}}\ \ (KMnO_4\ \text{solution})$$

6.87 To solve problems involving dilution, one must keep firmly in in mind that the amount of water in a solution does not change the quantity of solute present. (The original solution is designated soln(1) and the diluted solution as soln(2).

$$?\ \text{normality (1)} = ?\ \text{eq HCl} = 1^*\text{liter soln(1)} \times \frac{0.050\ \text{liter soln(2)}}{0.010\ \text{liter soln(1)}} \times$$

$$\frac{0.0410\ \text{liter NaOH soln}}{0.0050^\bullet\ \text{liter soln(2)}} \times \frac{0.255\ \text{eq NaOH}}{1^*\text{liter NaOH soln}} \times \frac{1^*\text{eq HCl}}{1^*\text{eq NaOH}} = \underline{\underline{10\ N}}$$

6.88 To determine the per cent ascorbic acid in the sample, the mass of ascorbic acid in the sample is determined and then the mass which would be present in a 100 g sample is calculated. (To make the problem set-up more compact, (A.A.) will be used for ascorbic acid.)

$$?\%(A.A.) = ?g(A.A.) = 100^{*}\cancel{g(s)} \times \frac{15.2^{\bullet}\cancel{ml\ soln}}{0.1000\cancel{g\ (s)}} \times \frac{0.0200\cancel{mol\ NaOH}}{10^{3}{}^{*}\cancel{ml\ soln}} \times$$

$$\frac{1^{*}\cancel{mol(A.A.)}}{2^{*}\cancel{mol\ NaOH}^{\Delta}} \times \frac{176.1\ g(A.A.)}{1^{*}\cancel{mol\ (A.A.)}} = \underline{\underline{26.8\ \%\ (A.A.)}}$$

$^{\Delta}$ascorbic acid is diprotic

6.89 Although there are many mixtures of the two acids which would have a weight of 0.1000 g, only one specific mixture of that weight would require exactly 20.4 ml of 0.0500 M NaOH. Lactic acid will be represented as (H-L) and caproic acid as (H-C).

let X = g(H-L) and Y = g(H-C). equivalent weights: (H-L) 90.080
(H-C) 116.16

$X + Y = 0.1000 \therefore Y = 0.1000 - X$ Since equivalents of acid are equal to equivalents of base these two quantities can be equated as shown below to give a second equation:

$$X\cancel{g(H\text{-}L)} \times \frac{1^{*}eq(H\text{-}L)}{90.080\cancel{g(H\text{-}L)}} + Y\cancel{g(H\text{-}C)} \times \frac{1^{*}eq(H\text{-}C)}{116.16\cancel{g(H\text{-}C)}} = \text{EQUIVALENTS}$$

$$\text{OF ACID} = \text{EQUIVALENTS OF BASE} = 20.4^{\bullet}\cancel{ml\ soln} \times \frac{0.0500\ eq\ NaOH}{10^{3}{}^{*}\cancel{ml\ soln}}$$

$^{\square}$precise to only 3 s.f.

$X/90.080 + Y/116.16 = 0.00102$ multiplying both sides by 116.16 gives:

$1.2895X + Y = 0.11848^{\square}$ substituting Y = 0.1000 - X (obtained above):

$1.2895X + 0.1000 - X = 0.11848^{\square} \therefore 0.2895X = 0.01848^{\Delta}$ $X = \underline{0.0638}$

$\therefore$ $\underline{\underline{0.064g^{\Delta}\text{ lactic acid}}}$ $\underline{\underline{0.036g^{\Delta}\text{ caproic acid}}}$ $^{\Delta}$precise to only 2 s.f.

6.90 There is only one unique mixture of MgO and CaO which will weigh 2.000 g and require the precise amount of HCl specified in this problem. First the net amount of HCl required for neutralization should be determined.

$$? \text{ mol HCl(net)} = 100^{*}\cancel{\text{ml soln}} \times \frac{1^{*}\text{mol HCl}}{10^{3*}\cancel{\text{ml soln}}} - 19.6\cancel{\text{ml soln(2)}} \times \frac{1^{*}\text{mol NaOH}}{10^{3*}\cancel{\text{soln}}}$$

$= \underline{0.0804 \text{ mol HCl(net)}}$ Second, the moles of oxide which would react with the net moles of HCl are determined.

$$? \text{ mol (MgO + CaO)} = 0.0804\cancel{\text{mol HCl}} \times \frac{1^{*}\text{mol(MgO + CaO)}}{2^{*}\cancel{\text{mol HCl}}} = \underline{0.0402 \text{ mol(oxide)}}$$

Third, let X equal the %MgO in the sample.(100 - X = %CaO) The equation below equates the sum of moles MgO and moles CaO to total moles of oxide sample.

$$2.000\cancel{\text{g}} \times \frac{X}{100^{*}}\cancel{\text{MgO}} \times \frac{1^{*}\text{mol MgO}}{40.31\cancel{\text{g MgO}}} + 2.000\cancel{\text{g}} \times \frac{(100^{*}X)}{100^{*}}\cancel{\text{CaO}} \times \frac{1^{*}\text{mol CaO}}{56.08\cancel{\text{g CaO}}} =$$

$0.0402 \text{ mol(oxide)}$ $\quad 4.96 \times 10^{-4}X + 3.57 \times 10^{-2} - 3.57 \times 10^{-4}X = 0.0402$

$1.39 \times 10^{-4}X = 0.0045$ $\quad \underline{X = 32} \therefore \underline{32\% \text{ MgO}} \quad \underline{68\% \text{ CaO}}$ In a 100 g sample of the oxide 32g would be MgO and 68g CaO. The corresponding weights of $MgCO_3$ and $CaCO_3$ are calculated below.

$$? \text{ g } MgCO_3 = 32\cancel{\text{g MgO}} \times \frac{84.32 \text{ g } MgCO_3}{40.31\cancel{\text{g MgO}}} = \underline{66.9 \text{ g } MgCo_3} \qquad ? \text{ g } CaCO_3 =$$

$$68\cancel{\text{g CaO}} \times \frac{100.1 \text{ g } CaCO_3}{56.08\cancel{\text{g CaO}}} = \underline{121 \text{ g } CaCO_3} \therefore \text{total} = 188 \text{ g}$$

$\% MgCO_3 = 66.9/188 \times 100 = \underline{\underline{36\% MgCO_3}} \therefore \underline{\underline{64\% CaCO_3}}$ (These are the % in the original carbonate mixture)

6.91 $\underline{CaCO_3 \longrightarrow CaO + CO_2}$ $\underline{CaO + H_2O \longrightarrow Ca(OH)_2}$

$\underline{Ca(OH)_2 + 2\ HCl \longrightarrow CaCl_2 + 2\ H_2O}$

First the mass of CaO in the 0.2000 g sample is determined from the mol HCl used for the titration. NOTE - The second and third equation above show a mole ratio for CaO - HCl of 1:2. (s) = original sample

$$?\ g\ CaO = 30.3^{*}\cancel{ml\ soln} \times \frac{0.1000\ \cancel{mol\ HCl}}{10^3\ {}^{*}\cancel{ml\ soln}} \times \frac{1^{*}\cancel{mol\ CaO}}{2^{*}\cancel{mol\ HCl}} \times \frac{56.08\ g\ CaO}{1^{*}\cancel{mol\ CaO}} =$$

$\underline{0.0850\ g\ CaO}$ $\therefore$ the other portion of the sample = $\underline{0.1150\ g}$

Second, the mass of $CaCO_3$ which decomposed to 0.0850 g CaO is calculated and with this value and the 0.1150 g of other material in the sample (as determined above) the % $CaCO_3$ in the original sample (s) may be determined.

$$?\ g\ CaCO_3 = 0.0850\ \cancel{g\ CaO} \times \frac{100.1\ g\ CaCO_3}{56.08\ \cancel{g\ CaO}} = \underline{0.1517\ g\ CaCO_3}$$

$$?\ \%\ CaCO_3 = ?\ g\ CaCO_3 = 100^{*}\cancel{g(s)} \times \frac{0.1517\ g\ CaCO_3}{(0.1517 + 0.1150)\ \cancel{g(s)}} = \underline{\underline{56.9\%\ CaCO_3}}$$

6.92 $8H_2SO_4 + 10FeSO_4 + 2KMnO_4 \longrightarrow 5Fe_2(SO_4)_3 + 2MnSO_4 + K_2SO_4 + 8H_2O$

$$?mol\ Fe^{2+} = 15.8\ \cancel{ml^{*}\ soln} \times \frac{0.00400\ \cancel{mol\ KMnO_4}}{10^3\ {}^{*}\cancel{ml\ soln}} \times \frac{10^{*}mol\ Fe^{2+}}{2^{*}\cancel{mol\ KMnO_4}} = \underline{0.000316\ mol\ Fe^{2+}}$$

$$?g\ FeSO_4 = 0.000316\ \cancel{mol\ Fe^{2+}} \times \frac{151.9\ g\ FeSO_4}{1^{*}\cancel{mol\ Fe^{2+}}} = \underline{0.0480\ g\ FeSO_4}$$ (cont.)

(6.92 continued) $?\% = ?gFeSO_4 = 100^* \not{g(s)} \times \frac{0.0480 \text{ g } FeSO_4}{0.1000 \not{g (s)}} = \underline{\underline{48.0\% \ FeSO_4}}$

6.93 | Let X = volume of 18.0 M soln $V_1 \times M_1 = V_2 \times M_2$

$X \bullet 18.0 = (100^* + X) \bullet 5.0^\bullet$ $(18.0 - 5.0) \bullet X = 500$ $X = \underline{\underline{38 \text{ ml}}}$

6.94 | Let X = ml of 1.00 M HCl M = mol/liter $\therefore$ $\underline{\text{mol = liter} \bullet \text{M}}$

$$0.600 \text{ M} = \frac{0.0500 \text{ liter}(2) \times 0.500 \text{ M} + X/1000^* \times 1.00 \text{ M}}{(0.0500 + X/1000) \text{ liter}}$$

$0.0300 + 6.00 \times 10^{-4}X = 0.025 + 10^{-3*}X$ $4.00 \times 10^{-4}X = 0.0050$

$X = \underline{\underline{12\text{ml } 1.00 \text{ M HCl}}}$

6.95 | (see 6.93) ? ml conc. $NH_3 = 250^*\text{ml dilute } NH_3 \times \frac{0.500^\bullet M \ NH_3}{14.8 \ M \ NH_3}$

$= \underline{\underline{8.45 \text{ ml conc } NH_3}}$

6.96 | (see 6.93) ? ml conc $H_2SO_4 = 400 \text{ ml dil } H_2SO_4 \times \frac{3.0 \ M \ H_2SO_4}{18.0 \ M \ H_2SO_4} =$

$\underline{\underline{67 \text{ ml conc } H_2SO_4}}$

6.97 (see 6.93) $? \text{ ml(f)} = 100^{*}\text{ml(i)} \times \dfrac{0.500^{\bullet}\text{mol}}{10^{3*}\text{ml(i)}} \times \dfrac{10^{3*}\text{ml(f)}}{0.200\text{ mol}} = \underline{\underline{250\text{ml(f}}}$

6.98 $? \text{ml } K_2Cr_2O_7 = 120^{*}\text{ml } H_2C_2O_4 \times \dfrac{0.850^{\bullet}\text{eq } H_2C_2O_4}{10^{3*}\text{ml } H_2C_2O_4} \times \dfrac{1^{*}\text{eq } K_2Cr_2O_7}{1^{*}\text{eq } H_2C_2O_4} \times$

$$\frac{10^{3*}\text{ml } K_2Cr_2O_7}{0.500\text{eq } K_2Cr_2O_7} = \underline{\underline{204 \text{ ml } K_2Cr_2O_7}}$$

6.99 $? \text{ml(f)} = 85.0^{\bullet}\text{ml(i)} \times \dfrac{1^{*}\text{N (i)}}{0.650 \text{ N (f)}} = \underline{131 \text{ ml(f)}}$

$? \text{ ml } H_2O \text{ added} = \text{ml(f)} - \text{ml(i)} = 131 - 85 = \underline{\underline{46 \text{ ml } H_2O \text{ added}}}$

6.100 The first step is the determination of the molarity of the the final solution.

$$? \text{ M(f)} = \frac{15.0 \text{ ml NaOH soln}}{0.025 \text{ liter (f)}} \times \frac{0.750 \text{ mol NaOH}}{10^{3*}\text{ml NaOH soln}} \times \frac{1^{*}\text{mol } H_2SO_4}{2^{*}\text{mol NaOH}} = \underline{0.225 \text{ M}}$$

The second step is to determine the volume of 0.225 M solution which would be produced by dilution of 250 ml of 1.40 M H_2SO_4.

$$? \text{ ml(f)} = 250^{*}\text{ml(i)} \times \frac{1.40^{\bullet}\text{M(i)}}{0.225 \text{ M(f)}} = \underline{\underline{1560 \text{ ml (0.225 M } H_2SO_4}}$$

PROBLEM SOLUTIONS FOR CHAPTER 7 OF BRADY & HUMISTON 3rd Ed.

7.29 (a) $? \text{ torr} = 1.50^{\bullet} \cancel{\text{atm}} \times \dfrac{760^{*} \text{ torr}}{1^{*} \cancel{\text{atm}}} = \underline{\underline{1140 \text{ torr}}}$

(b) $? \text{ atm} = 785^{\bullet} \cancel{\text{torr}} \times \dfrac{1^{*} \text{ atm}}{760^{*} \cancel{\text{torr}}} = \underline{\underline{1.03 \text{ atm}}}$

(c) $? \text{ Pa} = 3.45^{\bullet} \cancel{\text{atm}} \times \dfrac{101{,}325 \text{ Pa}}{1 \cancel{\text{atm}}} = \underline{\underline{3.50 \times 10^{5} \text{ Pa}}}$

(d) $? \text{ kPa} = 3.45^{\bullet} \cancel{\text{atm}} \times \dfrac{101.325 \text{ kPa}}{1 \cancel{\text{atm}}} = \underline{\underline{350 \text{ kPa}}}$

(e) $? \text{ Pa} = 165^{\bullet} \cancel{\text{torr}} \times \dfrac{101{,}325 \text{ Pa}}{760^{*} \cancel{\text{torr}}} = \underline{\underline{2.20 \times 10^{4} \text{ Pa}}}$

(f) $? \text{ atm} = 342^{\bullet} \cancel{\text{kPa}} \times \dfrac{1^{*} \text{ atm}}{101.325 \cancel{\text{kPa}}} = \underline{\underline{3.38 \text{ atm}}}$

(g) $? \text{ torr} = 11.5^{\bullet} \cancel{\text{kPa}} \times \dfrac{760^{*} \text{ torr}}{101.325 \cancel{\text{kPa}}} = \underline{\underline{86.3 \text{ torr}}}$

7.30 $? \text{ t} = 15.8^{\bullet} \cancel{\text{cm Hg}} \times \dfrac{760^{*} \text{ torr}}{76^{*} \cancel{\text{cm Hg}}} = \underline{\underline{158 \text{ torr}}}$

7.31 $P = [733 \text{ torr} + 65 \text{ torr}]^{\bullet} = \underline{\underline{798 \text{ torr}}}$

7.32 $? \text{ mm Hg} = (774 \cancel{\text{torr}} - 535 \cancel{\text{torr}})^{\bullet} \times \dfrac{1^{*} \text{ mm Hg}}{1^{*} \cancel{\text{torr}}} = \underline{\underline{239 \text{ mm Hg}}}$

7.33 $? \text{ torr} = (755 \cancel{\text{mm Hg}} + 17 \cancel{\text{mm Hg}})^{\bullet} \times \dfrac{1^{*} \text{ torr}}{1^{*} \cancel{\text{mm Hg}}} = \underline{\underline{772 \text{ torr}}}$

7.34 $? \text{ torr (B)} = 836 \text{ torr (A)} + 74 \cancel{\text{cm(o)}} \times \dfrac{0.847^{\bullet} \cancel{\text{cm } H_2O}}{1^{\bullet} \cancel{\text{cm(o)}}} \times$

$\dfrac{1^{*} \cancel{\text{cm Hg}}}{13.55 \cancel{\text{cm } H_2O}} \times \dfrac{10^{*} \text{ torr}}{1^{*} \cancel{\text{cm Hg}}} = \underline{\underline{882 \text{ torr (B)}}}$

7.35 The answer to the question is <u>NO</u>! A pressure of 1 atm is equa to 33.9 ft of water. Even if the pump had perfect seals, it could not raise water 35 ft by suction.

7.36

	(I) $\longrightarrow$	(F)
P	740 torr	900 torr
V	350 ml	? ml

$? \text{ ml(F)} = 350 \text{ ml(I)} \times \dfrac{740 \cancel{\text{torr}}}{900 \cancel{\text{torr}}} = \underline{\underline{288 \text{ ml(F)}}}$

7.37

'liter' will be abbreviated 'ltr' to make set-ups more compact.

	(I) →	(F)
P	2.75 atm	800 torr
V	1.45 ltr	? ltr

$$? \text{ ltr(F)} = 1.45^{\bullet} \text{ltr(I)} \times \frac{2.75 \cancel{\text{atm}}}{800 \cancel{\text{torr}}} \times \frac{760 \cancel{\text{torr}}}{1^{*} \cancel{\text{atm}}} = \underline{\underline{3.79 \text{ ltr(F)}}}$$

7.38

	(I) →	(F)
P	475 torr	? torr
V	540 ml	320 ml

$$? \text{torr(F)} = 475 \text{ torr(I)} \times \frac{540 \cancel{\text{ml}}}{320 \cancel{\text{ml}}} = \underline{\underline{802 \text{ torr(F)}}}$$

7.39

	(I) →	(F)
P	20.0 psi	? psi
V	35.0 ft^3	40.0 ft^3

$$? \text{psi(F)} = 20.0 \text{ psi(I)} \times \frac{35.0 \cancel{ft^3}}{40.0 \cancel{ft^3}} = \underline{\underline{17.5 \text{ psi(F)}}}$$

7.40

The volume of a cylinder equals the cross sectional area (πr^2) multiplied by the length.

et X = length of the down stroke $\quad V_i = 75.0 \text{ cm} \bullet \pi r^2, \; V_f = (75.0-X)\text{cm}\pi r^2$

$$_fV_f = P_iV_i \qquad V_f = V_i \bullet \frac{P_i}{P_f} \qquad (75.0-X)\cancel{\text{cm}\bullet\pi r^2} = 75.0 \cancel{\text{cm}\bullet\pi r^2} \bullet \frac{1.00 \cancel{\text{atm}}}{5.50 \cancel{\text{atm}}}$$

$$5.0 - X = \frac{75.0}{5.50} = 13.64 \qquad X = 75.0 - 13.64 = \underline{\underline{61.4 \text{ cm (downstroke)}}}$$

7.41

(I) → (F)

	(I)	(F)
P	1*atm	1*atm
V	1.50 ltr	? ltr
T	298 K	373 K

$$?\text{ltr(F)} = 1.50\ \text{ltr(I)} \times \frac{373\ \cancel{K}}{298\ \cancel{K}} = \underline{\underline{1.88\ \text{ltr(F)}}}$$

7.42

(I) → (F)

	(I)	(F)
V	2.0 ltr	? ltr
T	298 K	244.1 K

$$?\text{ltr(F)} = 2.0^{\bullet}\text{ltr(I)} \times \frac{244.1\ \cancel{K}}{298\ \cancel{K}} = \underline{\underline{1.61\text{t (F)}}}$$

7.43

(I) → (F)

	(I)	(F)
V	2.00 ltr	? ltr
T	299 K	373 K

$$?\text{ltr(F)} = 2.00^{\bullet}\text{ltr(I)} \times \frac{373\ \cancel{K}}{299\ \cancel{K}} = \underline{\underline{2.49\ \text{l (F)}}}$$

7.44

(I) → (F)

	(I)	(F)
P	750 torr	? torr
T	293 K	313 K

$$?\text{torr(F)} = 350\text{torr(I)} \times \frac{313\ \cancel{K}}{293\ \cancel{K}} = \underline{\underline{374\text{tor (F)}}}$$

7.45

(I) → (F)

	(I)	(F)
P	655 torr	825 torr
T	(273+25) K	? K

$$?\ \text{K(F)} = 298\ \text{K(I)} \times \frac{825\ \cancel{\text{torr}}}{655\ \cancel{\text{torr}}} = \underline{375\ \text{K(F}}$$

$\underline{\underline{102^{\circ}\text{C}}}$

7.46

	(I) →	(F)
P	43.7 psi	? psi
T	273+18.3 K	273+54.4 K

29 psi(G) = (29+14.7) psi

65°F = 18.3°C   130°F = 54.4°C

$$? \text{ psi(F)} = 43.7^{\bullet}\text{psi(I)} \times \frac{327.4 \text{ K}}{291.3 \text{ K}} = \underline{49.1 \text{ psi}} = \underline{\underline{34 \text{ psi(G)}}}$$

7.47

	(I) →	(F)
V	285 ml	350 ml
T	298 K	? K

$$?\text{K(F)} = 298 \text{ K(I)} \times \frac{350 \text{ ml}}{285 \text{ ml}} = \underline{366 \text{ K(F)}}$$

$\underline{\underline{93^\circ\text{C}}}$

7.48

	(I) →	(F)
P	1*atm	1*atm
V	0.400 ltr	0.850 ltr
T	305 K	? K

$$?\text{K(F)} = 305 \text{ K(I)} \times \frac{0.850 \text{ ltr}}{0.400 \text{ ltr}} = \underline{648 \text{ K(F)}}$$

$\underline{\underline{375^\circ\text{C}}}$

7.49

	(I) →	(F)
P	645 torr	? torr
V	50 ml	65 ml
T	298 K	308 K

(a) $?\text{torr(F)} = 645\text{torr(I)} \times \frac{50^{\bullet}\text{ ml}}{65 \text{ ml}} = \underline{\underline{496 \text{ torr}}}$

(NO TEMPERATURE CHANGE)

(b) $$? \text{ torr(F)} = 645 \text{ torr(I)} \times \frac{50 \text{ ml}}{65 \text{ ml}} \times \frac{308 \text{ K}}{298 \text{ K}} = \underline{\underline{513 \text{ torr(F)}}}$$

7.50 (I) ⟶ (F)

	(I)	(F)
P	450 torr	? torr
V	300 ml	200 ml
T	300K	293 K

$$?\text{torr(F)} = 450\text{torr(I)} \times \frac{300 \cancel{\text{ml}}}{200 \cancel{\text{ml}}} \times \frac{293 \cancel{K}}{300 \cancel{K}}$$

$$= \underline{\underline{659 \text{ torr(F)}}}$$

7.51 (I) ⟶ (F)

	(I)	(F)
P	700 torr	585 torr
V	2.00 ltr	5.00 ltr
T	298 K	? K

$$?\text{K(F)} = 298 \text{ K(I)} \times \frac{585 \cancel{\text{torr}}}{700 \cancel{\text{torr}}} \times \frac{5.00 \cancel{\text{ltr}}}{2.00 \cancel{\text{ltr}}}$$

$$= \underline{623 \text{ K(F)}} \qquad \underline{\underline{350^\circ C}}$$

7.52 (I) ⟶ (F)

	(I)	(F)
P	760 torr	650 torr
V	1.00 ltr	1.00 ltr
T	273 K	298 K
N	1.96 g	? g

At the initial conditions, 1.00 liter of CO_2 weighs 1.96 g. To determine the density of the gas at the final conditions we determine the weight of one liter of CO_2 at the final conditions. (SEE NOTES AT THE START OF CHAPTER 7 OF THIS MANUAL.)

$$\text{density} = ?\ \text{g/ltr} = ?\ \text{g(F)} = 1.96 \text{ g(I)} \times \frac{650 \cancel{\text{torr}}}{760 \cancel{\text{torr}}} \times \frac{273 \cancel{K}}{298 \cancel{K}} = \underline{1.54 \text{ g(F)}}$$

$$\text{DENSITY} = \underline{\underline{1.54 \text{ g/ltr}}}$$

7.53 (I) ⟶ (STP)

	(I)	(STP)
P	450 torr	760 torr
V	50.0 ml	? ml
T	308 K	273 K

$$?\text{ml(STP)} = 50.0\text{ml(I)} \times \frac{450 \cancel{\text{torr}}}{760 \cancel{\text{torr}}} \times \frac{273 \cancel{K}}{308 \cancel{K}}$$

$$= \underline{\underline{26.2 \text{ ml(STP)}}}$$

7.54 200 torr + 500 torr + 150 torr = 850 torr

7.55 total pressure = $300 \text{ torr} \times \frac{2.00 \cancel{\text{ltr}}}{1.00 \cancel{\text{ltr}}} + 80 \text{ torr} \times \frac{2.00 \cancel{\text{ltr}}}{1.00 \cancel{\text{ltr}}} =$ 760 torr = 1.00 atm

7.56

	(I) →	(F)
P	740 torr	? torr
V	20.0 ml	50.0 ml

$?\text{torr(F)} = 740 \text{ torr(I)} \frac{20.0 \cancel{\text{ml}}}{50.0 \cancel{\text{ml}}} = 296 \text{ torr(F)}$ (NITROGEN)

	(I) →	(F)
P	640 torr	? torr
V	30.0 ml	50.0 ml

$?\text{torr(F)} = 640 \text{ torr(I)} \times \frac{30.0 \cancel{\text{ml}}}{50.0 \cancel{\text{ml}}} = 384 \text{ torr(F)}$ (OXYGEN)

$P_{total} = P_{N_2} + P_{O_2} = 296 \text{ torr} + 384 \text{ torr} = 680 \text{ torr}$

7.57 The solution of this problem involves several steps. First, the partial volume of oxygen in the final mixture is determined and this value is subtracted from the total volume of 100ml to obtain the partial volume of nitrogen.

	(I) →	(F)
P	400 torr	800 torr
V	50 ml	? ml(O_2)
T	333 K	323 K

$? \text{ l(F)} = 50 \text{ ml(I)} \times \frac{400 \cancel{\text{torr}}}{800 \cancel{\text{torr}}} \times \frac{323 \cancel{\text{K}}}{333 \cancel{\text{K}}} =$

24 ml(F) (O_2) ∴ 76 ml(F) (N_2)

Second, the original volume of nitrogen which would yield 76ml(F)(N_2) is calculated.

(7.57 continued)

	(F)	(I)
P	800 torr	400 torr
V	76 ml	? ml (N_2)
T	323 K	313 K

$?ml(I) = 76 ml(F) \times \frac{800 \text{ torr}}{400 \text{ torr}} \times \frac{313 K}{323 K} =$

150 ml(I) (only 2 s.f.)

7.58

	(I)	(STP)
P	(700-24) torr	760 torr
V	100 ml	? ml
T	298 K	273 K

$?ml(STP) = 100 ml(I) \times \frac{676 \text{ torr}}{760 \text{ torr}} \times \frac{273 K}{298 K}$

= 81.5 ml (STP)

7.59 The first step is to determine the total number of moles of gas present in the vessel. PV=nRT

$n = PV/RT = 720/760 \text{ atm} \cdot 0.200 \text{ ltr} / 0.0821 \text{ ltr} \cdot \text{atm} \cdot \text{mol}^{-1} K^{-1} \cdot (273+35) K$

$n = 7.49 \times 10^{-3}$ mol   mole fraction (N_2) = mol(N_2) / total mol

$X_{N_2} = 2.0 \times 10^{-3} / 7.49 \times 10^{-3}$ = 0.27 (mole fraction N_2) (a)

$P = nRT/V = 0.0020 \text{ mol} \cdot 0.0821 \text{ ltr} \cdot \text{atm} \cdot \text{mol}^{-1} \cdot K^{-1} \cdot (273+35) K / 0.200 \text{ ltr}$

P = 0.25 atm

(b) 190 torr

(c) $P_{O_2} = P_{total} - P_{N_2}$ = 720 torr - 190 torr

P_{O_2} = 530 torr

$n = PV / RT = 530/760 \text{ atm} \cdot 0.200 \text{ ltr} / 0.0821 \text{ ltr} \cdot \text{atm} \cdot \text{mol}^{-1} \cdot K^{-1} \cdot (273+35) K$

n = 0.0055 mol (O_2) (d)

7.60

$$\underline{X}_{N_2} = \frac{569 \text{ torr}}{569 \text{ torr} + 116 \text{ torr} + 28 \text{ torr} + 47 \text{ torr}} = \underline{\underline{0.749}} \text{ (mole fraction)}$$

$$\underline{X}_{O_2} = \frac{116 \text{ torr}}{760 \text{ torr}} = \underline{\underline{0.153}} \text{ (mole fraction)} \qquad \underline{X}_{CO_2} = \frac{28 \text{ torr}}{760 \text{ torr}} = \underline{\underline{0.037}} \text{ (mol fr)}$$

$$\underline{X}_{H_2O} = 47 \text{ torr} / 760 \text{ torr} = \underline{\underline{0.062}} \text{ (mole fraction)}$$

7.61

The sum of the partial pressures equals the total pressure.
$P_{(F)}(CO_2)$ = 900 torr − 800 torr = <u>100 torr</u>

(F) ⟶ (I)

	(F)	(I)
P	100 torr	700 torr
V	500 ml	? ml
T	293 K	303 K

$$?\text{ml(I)} = 500\text{ml(F)} \times \frac{100 \text{ torr}}{700 \text{ torr}} \times \frac{303 \text{ K}}{293 \text{ K}} =$$

$\underline{\underline{73.9 \text{ ml(I) } (CO_2)}}$

7.62

(a) $V = nRT/P = 0.200 \text{ mol} \cdot 0.0821 \text{ ltr} \cdot \text{atm} \cdot \text{mol}^{-1} \cdot \text{K}^{-1} \cdot 273\text{K} / 1 \text{ atm}$

$V = \underline{\underline{4.48 \text{ ltr}}}$

(b) $$V = \frac{12.4/70.9 \text{ mol} \cdot 0.0821 \text{ ltr} \cdot \text{atm} \cdot \text{mol}^{-1} \cdot \text{K}^{-1} \cdot 273\text{K}}{1 \text{ atm}} =$$

$V = \underline{\underline{3.92 \text{ ltr}}}$

(c) total mol = 0.150 ∴ $V = 0.150 \cdot 0.0821 \cdot 273 / 1 = \underline{\underline{3.36 \text{ ltr}}}$

7.63

$$?\text{g}SO_2\text{(STP)} = 0.245 \text{ ltr } SO_2\text{(STP)} \times \frac{64.1 \text{ g } SO_2}{22.4 \text{ ltr } SO_2\text{(STP)}} = \underline{\underline{0.701 \text{ g } SO_2}}$$

7.64

$d = \text{mass/volume} = 58.1 \text{ g} / 22.4 \text{ ltr (STP)} = \underline{\underline{2.59 \text{ g/ltr}}}$

7.65 $?\text{MW} = ?\ \text{g} = 1^*\text{mol} = 22.4\ \cancel{\text{ltr(STP)}} \times 1.96\ \text{g} / 1^*\cancel{\text{ltr(STP)}} = \underline{\underline{43.9\ \text{g}}}$

7.66 $$\frac{?\ \text{Pa}\bullet\text{m}^3}{\text{mol}\bullet\text{K}} = \frac{0.0821\ \cancel{\text{ltr}}\bullet\cancel{\text{atm}}}{1^*\text{mol}\bullet\text{K}} \times \frac{10^3{}^*\cancel{\text{cm}^3}}{1^*\cancel{\text{ltr}}} \times \frac{1^*\text{m}^3}{(10^2\ \cancel{\text{cm}})^3} \times \frac{1.013\bullet10^5\ \text{Pa}}{1^*\cancel{\text{atm}}} =$$

$\underline{\underline{8.31\ \ \text{Pa}\bullet\text{m}^3/\text{mol}\bullet\text{K}}}$

7.67 $n = PV/RT = 1^*\cancel{\text{atm}}\bullet0.250\ \cancel{\text{ltr}} \,/\, 0.0821\ \cancel{\text{ltr}}\bullet\cancel{\text{atm}}\bullet\text{mol}^{-1}\bullet\cancel{\text{K}^{-1}}\bullet298\ \cancel{\text{K}} =$

$\underline{0.0102\ \text{mol}}$ $\qquad \text{MW} = ?\ \text{g} = 1^*\cancel{\text{mol}} \times 0.164\ \text{g}/0.0102\ \cancel{\text{mol}} = \underline{\underline{16.0\ \text{g}}}$

7.68 $P = nRT/V$ $$?\text{atm} = 25{,}000\ \cancel{\text{g}} \times \frac{1^*\cancel{\text{mol}}}{18.0\ \cancel{\text{g}}} \times \frac{0.0821\ \cancel{\text{ltr}}\bullet\text{atm}}{1^*\cancel{\text{mol}}\bullet\cancel{\text{K}}} \times \frac{473\ \cancel{\text{K}}}{10^3{}^*\cancel{\text{ltr}}}$$

$\underline{\underline{54\ \text{atm}}}$ $\quad \underline{\underline{4.1 \times 10^4\,\text{torr}}}$

7.69 $$n = \frac{1^*\cancel{\text{atm}}\bullet1^*\cancel{\text{ltr}}}{303\ \cancel{\text{K}}} \times \frac{1^*\text{mol}\bullet\cancel{\text{K}}}{0.0821\ \cancel{\text{ltr}}\bullet\cancel{\text{atm}}} = \underline{0.0402\ \cancel{\text{mol}}} \times \frac{1.81\ \text{g}}{0.0402\ \cancel{\text{mol}}} = \underline{\underline{45.0\ \text{g}}}$$

7.70 $$V = \frac{nRT}{P} = \frac{0.234\ \cancel{\text{g}} \times 0.0821\ \text{ltr}\bullet\cancel{\text{atm}}\bullet303\ \cancel{\text{K}}}{17.03\ \cancel{\text{g}}\,\cancel{\text{mol}^{-1}}\bullet{}^*\cancel{\text{mol}}\bullet\cancel{\text{K}} \times 0.847\ \cancel{\text{atm}}} = \underline{0.403\ \text{ltr}} \quad \underline{\underline{403\ \text{ml}}}$$

7.71 $$?\ \text{mol C} = 80.0\ \cancel{\text{g C}} \times \frac{1^*\text{mol C}}{12.0\ \cancel{\text{g C}}} = \underline{6.67\ \text{mol C}}$$

(7.71 continued)(a) $? \text{ mol H} = 20.0 \cancel{\text{g H}} \times \frac{1^* \text{mol H}}{1.008 \cancel{\text{g H}}} = \underline{19.8 \text{ mol H}}$ $\therefore$ $\underline{\underline{CH_3}}$

(b) $n = \frac{PV}{RT} = \frac{1^* \cancel{\text{atm}} \times 0.500 \cancel{\text{ltr}}}{0.0821 \cancel{\text{ltr}} \cdot \cancel{\text{atm}} \cdot \text{mol}^{-1} \cdot \cancel{K^{-1}} \cdot 273 \cancel{K}} = \underline{0.0223 \text{ mol}}$

$?MW = ? \text{ g} = 1^* \cancel{\text{mol}} \times \frac{0.6695 \text{ g}}{0.0223 \cancel{\text{mol}}} = \underline{\underline{30.0 \text{ g}}}$ (c) $\underline{\underline{C_2H_6}}$

7.72 (a) $?\% \text{ C} = ?\text{g C} = 100^* \cancel{\text{g(s)}} \times \frac{0.482 \cancel{\text{g } CO_2}}{0.2000 \cancel{\text{g(s)}}} \times \frac{12.01 \text{g C}}{44.01 \cancel{\text{g } CO_2}} = \underline{\underline{65.8\% \text{ C}}}$

$?\% \text{ H} = ?\text{g H} = 100^* \cancel{\text{g(s)}} \times \frac{0.271 \cancel{\text{g } H_2O}}{0.2000 \cancel{\text{g(s)}}} \times \frac{2.016 \text{g H}}{18.02 \cancel{\text{g } H_2O}} = \underline{\underline{15.2\% \text{ H}}}$

$n = \frac{PV}{RT}$

$n = \frac{755/760 \cancel{\text{atm}} \cdot 0.0423 \cancel{\text{ltr}}}{0.0821 \cancel{\text{ltr}} \cdot \cancel{\text{atm}} \cdot \text{mol}^{-1} \cdot \cancel{K^{-1}} \cdot 299.5 \cancel{K}} = \underline{0.00171 \text{ mol } N_2}$ (mole N_2 from the 0.2500 g sample)

$? \text{ mol N} = 0.2000 \cancel{\text{g(s)}} \times \frac{0.00171 \cancel{\text{mol } N_2}}{0.2500 \cancel{\text{g (s)}}} \times \frac{2^* \text{mol N}}{1^* \cancel{\text{mol } N_2}} = \underline{0.00274 \text{ mol N}}$ (mol N per a 0.2000g(s)

$?\% \text{ N} = ?\text{g N} = 100^* \cancel{\text{g(s)}} \times \frac{0.00274 \cancel{\text{mol N}}}{0.2000 \cancel{\text{g(s)}}} \times \frac{14.01 \text{g N}}{1^* \cancel{\text{mol N}}} = \underline{\underline{19.2\% \text{ N}}}$

(b) $? \text{ mol C} = 0.482^* \cancel{\text{g } CO_2} \times \frac{1^* \text{mol C}}{44.00 \cancel{\text{g } CO_2}} = \underline{0.0110 \text{ mol C}}$

(7.72 continued)

$$?mol\ H = 0.271 \bullet \cancel{g\ H_2O} \times \frac{2^*\ mol\ H}{18.02\ \cancel{g\ H_2O}} = \underline{0.0301\ mol\ H}$$

$$\frac{C_{0.0110}}{0.00274}\ \frac{H_{0.0301}}{0.00274}\ \frac{N_{0.00274}}{0.00274} = \underline{\underline{C_4H_{11}N}}$$

7.73

$$?\ mol\ NH_3 = \frac{750/760\ \cancel{atm} \bullet 0.120\ \cancel{ltr}}{0.0821\ \cancel{ltr \bullet atm} \bullet mol^{-1} \bullet \cancel{K}^{-1} \bullet 298\cancel{K}} = \underline{4.84 \times 10^{-3}\ mol\ NH_3}$$

$$?\ mol\ O_2 = \frac{635/760\ \cancel{atm} \bullet 0.165\ \cancel{ltr}}{0.0821\ \cancel{ltr \bullet atm} \bullet mol^{-1} \bullet \cancel{K}^{-1} \bullet 323\cancel{K}} = \underline{5.20 \times 10^{-3}\ mol\ O_2}$$

Since 4 mole NH_3 require 5 mole O_2 (balanced chemical equation) oxygen is the limiting reactant.

$$?\ mol\ NH_3(react) = 5.20 \times 10^{-3}\ \cancel{mol\ O_2} \times \frac{4^*\ mol\ NH_3}{5^*\ \cancel{mol\ O_2}} = \underline{4.16 \times 10^{-3} mol\ NH_3(r)}$$

$$?\ mol\ NH_3(XS) = 4.84 \times 10^{-3} - 4.16 \times 10^{-3} = \underline{6.8 \times 10^{-4}\ mol\ NH_3(XS)}$$

$$?\ total\ mol\ gas\ produced = 5.20 \times 10^{-3}\ \cancel{mol\ O_2} \times 10^*\ mol(g.p.)/5^*\ \cancel{mol\ O_2} =$$

$\underline{1.04 \times 10^{-2}\ mol(g.p.)}$ total mol gas after reaction $= 1.04 \times 10^{-2} +$

$6.8 \times 10^{-4} = \underline{1.11 \times 10^{-2} mol}$

$$P = \frac{nRT}{V} = \frac{1.11 \times 10^{-2}\ \cancel{mol} \bullet 0.0821\ \cancel{ltr} \bullet atm}{0.300\ \cancel{ltr}\ \ \cancel{mol} \bullet \cancel{K}} \times 423\ \cancel{K} = \underline{1.28\ atm} = \underline{\underline{975\ torr}}$$

7.74

$$?\text{ml } N_2(\text{STP}) = 400^*\text{ml } NH_3(\text{STP}) \times \frac{1^*\text{ml } N_2(\text{STP})}{2^*\text{ml } NH_3(\text{STP})} = \underline{\underline{200\text{ml } N_2(\text{STP})}}$$

$\underline{\underline{600\text{ml } H_2(\text{STP})}}$

7.75

(a) $$?\text{ml } N_2(\text{STP}) = 1.40^\bullet \times 10^{-3}\text{mol NO} \times \frac{1^*\text{mol } N_2}{2^*\text{mol NO}} \times \frac{22{,}400\text{ml } N_2(\text{STP})}{1^*\text{mol } N_2}$$

$= \underline{\underline{15.7\text{ml } N_2(\text{STP})}}$

(b) $$?\text{ml } N_2(\text{STP}) = 1.3^\bullet \times 10^{-3}\text{g } H_2 \times \frac{1^*\text{mol } H_2}{2.016\text{g } H_2} \times \frac{1^*\text{mol } N_2}{2^*\text{mol } H_2} \times \frac{22{,}400\text{ml } N_2(\text{STP})}{1^*\text{mol } N_2} = \underline{\underline{7.2\text{ml } N_2(\text{STP})}}$$

7.76

$$2\ KClO_3 \longrightarrow 2\ KCl + 3\ O_2 \qquad n = PV/RT$$

(a)

$$?\text{ mol} = \frac{(600-32)/760\text{ atm}\cdot 0.150\text{ ltr}}{0.0821\text{ ltr}\cdot\text{atm}\cdot\text{mol}^{-1}\cdot K^{-1}\cdot 303K} = \underline{4.51 \times 10^{-3}\text{mol } O_2} \times \frac{32.00\text{g } O_2}{1^*\text{mol } O_2}$$

$= \underline{\underline{0.144\text{ g } O_2}}$

(b) $$?\text{g } KClO_3 = 4.51 \times 10^{-3}\text{mol } O_2 \times \frac{2^*\text{mol } KClO_3}{3^*\text{mol } O_2} \times \frac{122.6\text{g} KClO_3}{1^*\text{mol} KClO_3}$$

$= \underline{\underline{0.368\text{g } KClO_3}}$

7.77

$$?\text{mol } NO_2 = 10.0^\bullet\text{g } HNO_3 \times \frac{1^*\text{mol } HNO_3}{63.0\text{g } HNO_3} \times \frac{3^*\text{mol } NO_2}{2^*\text{mol } HNO_3} = \underline{0.238\text{mol} NO_2}$$

$$V = \frac{nRT}{P} = \frac{0.238\text{ mol}\cdot 0.0821\text{ ltr}\cdot\text{atm}\cdot 298K}{770/760\text{ atm}\quad \text{mol}\cdot K} = \underline{5.75\text{ ltr}} \qquad \underline{\underline{5750\text{ ml}}}$$

7.78 $\frac{r_{He}}{r_{Ne}} = \left(\frac{M_{Ne}}{M_{He}}\right)^{\frac{1}{2}} = \left[\frac{20.183}{4.0026}\right]^{\frac{1}{2}} = \underline{\underline{2.25}}$ (faster for He)

7.79 at temperature "T": $\overline{KE}_{CH_4} = \overline{KE}_{CO_2}$ $\overline{KE} = 1/2\ m\bar{v}^2$

$$1/2\ m_{CH_4}\bar{v}^2_{CH_4} = 1/2\ m_{CO_2}\bar{v}^2_{CO_2} \quad \bar{v}^2_{CO_2} = \frac{m_{CH_4}\ \bar{v}^2_{CH_4}}{m_{CO_2}} \quad \bar{v}_{CO_2} = \left[\frac{m_{CH_4}}{m_{CO_2}}\right]^{1/2} \bullet \bar{v}_{CH_4}$$

$\bar{v}_{CH_4} = 1000\text{mph}$ $\therefore$ $\bar{v}_{CO_2} = (16/44)^{1/2} \times 1000\text{mph}$ $\bar{v}_{CO_2} = \underline{\underline{600\text{mph}}}$

7.80 $r_{un}/r_{NH_3} = 2.92 = (m_{NH_3}/m_{un})^{1/2}$ $m_{NH_3}/m_{un} = 8.53$

$8.53\ m_{un} = 17.0$ $m_{un} = \underline{\underline{1.99\ g/mol}}$

7.81 How much more precise answer can be obtained by the use of van der Waals equation? $(P + a/V^2)(V - b) = RT$

$$P = \frac{RT}{V-b} - \frac{a}{V^2} = \frac{0.082054\ \cancel{ltr}\bullet atm \bullet 273.18\cancel{K}}{\cancel{mol}\bullet\cancel{K}(22.400\cancel{ltr/mol} - 0.02370\ \cancel{ltr/mol})} -$$

$$\frac{0.034\ \cancel{ltr^2}\bullet atm/\cancel{mol^2}}{(22.400\ \cancel{ltr/mol})^2} = \underline{\underline{1.0017\ atm}}$$

For an ideal gas the pressure would equal 1.000 atm. In this example He behaves almost exactly as an I. G.

7.82 $(P + a/V^2)(V - b) = RT$

$$P = \frac{RT}{V-b} - \frac{a}{V^2} = \frac{mol \cdot 0.082054\ ltr \cdot atm \cdot 273.15\ K}{(22.400-0.06380)\ ltr \cdot mol \cdot K} - \frac{5.489\ ltr^2 \cdot atm}{(22.400\ ltr)^2} = \underline{\underline{0.9925\ atm}}$$

(ideal gas would show P = 1atm)

7.83

$$?psi = 1^{*} atm \times \frac{76.0 cm\ Hg}{1^{*} atm} \times \frac{13.6 g\ Hg}{1^{*} cm^3\ Hg} \times \left(\frac{2.54 cm}{1^{*} in}\right)^2 \times \frac{1^{*} lb\ Hg}{454 g\ Hg} = \underline{\underline{14.7\ lb\ in^{-2}}}$$

7.84 $PV = nRT$

$$V = \frac{nRT}{P} = \frac{(0.0244/32.00)\ mol \cdot 0.0821\ ltr \cdot atm \cdot 296 K}{(740 - 21)/760\ atm \cdot mol \cdot K}$$

$= \underline{0.0196\ ltr}$ $\quad \underline{\underline{19.6\ ml}}$

7.85 $PV = nRT$

$$n = \frac{PV}{RT} = \frac{(800/760)\ atm \cdot 10^{*} ltr}{0.0821\ ltr \cdot atm \cdot mol^{-1} \cdot K^{-1} \cdot 303 K} = \underline{\underline{0.423\ mol}}$$

(a) (TOTAL MOLE)

(b) $$?mol\ CO_2 = 8.0\ g\ CO_2 \times \frac{1^{*} mol\ CO_2}{44.01 g\ CO_2} = \underline{0.18\ mol\ CO_2}$$

$$?mol\ O_2 = 6.0\ g\ O_2 \times \frac{1^{*} mol\ O_2}{32.00 g\ O_2} = \underline{0.19\ mol\ O_2}$$

$$?X_{CO_2} = \frac{0.18}{0.423} = \underline{\underline{0.43}} \qquad ?X_{O_2} = \frac{0.19}{0.423} = \underline{\underline{0.45}} \qquad ?X_{N_2} = 1^{*} - 0.43 - 0.45 =$$

$\underline{\underline{0.12}}$

$$?mol\ N_2 = 0.433 (total\ mol) \times \frac{0.12\ mol\ N_2}{1^{*} (total\ mol)} = \underline{0.051 mol\ N_2}$$

(7.85 continued) (c) $P = \frac{nRT}{V} = \frac{0.18 \cancel{mol} \cdot 0.0821 \cancel{ltr} \cdot atm \cdot 303 \cancel{K}}{10 \cancel{ltr} \quad \cancel{mol} \cdot \cancel{K}} = \underline{0.45 \text{ atm}}$

(340 torr)(CO_2) 360 torr (O_2) 100 torr (N_2)

(d) $0.051 \cancel{mol\ N_2} \times \frac{28.01 g\ N_2}{1 \cancel{mol\ N_2}} = \underline{1.4 g\ N_2}$

7.86

(I) ⟶ (F)

	(I)	(F)
P	800-41torr	? torr
V	500 ml	250 ml
T	308 K	308 K

$?torr(F) = (800-41)torr(I) \times \frac{500 \cancel{ml}}{250 \cancel{ml}} =$

(only 3 s.f.)

1518 torr (dry) 1518+41 = 1560 torr (wet)

7.87

(S) ⟶ (F)

	(S)	(F)
P	760 torr	33.7 torr*
V	22.4 ltr	1.00 ltr
T	273 K	304 K
N	18.02 g	? g H_2O

*saturation V.P. H_2O at 31°C

$?g\ H_2O(F) = 18.02g\ H_2O(S) \times \frac{33.7 \cancel{torr} \cdot}{760 \cancel{torr} \cdot}$

$\frac{1.00 \cancel{ltr} \cdot 273 \cancel{K}}{22.4 \cancel{ltr} \cdot 304 \cancel{K}} = \underline{0.0320 \text{ g } H_2O}$

7.88

(I) ⟶ (STP)

	(I)	(STP)
P	743 torr*	760 torr
V	280 ml	? ml
T	293 K	273 K

$*P_g(\text{dry}) = (763 - 28.4 \cdot \frac{1^* torr}{13.55 mm\ H_2O} - 17.5)$

$?ml(STP) = 280\ ml(I) \times \frac{743 \cancel{torr} \cdot 273 \cancel{K}}{760 \cancel{torr} \cdot 293 \cancel{K}}$

$= \underline{255\ ml(STP)}$

7.89

$$? \text{ J} = 1^{*}\cancel{\text{ltr}\cdot\text{atm}} \times \frac{101{,}325\,\cancel{\text{Pa}}}{1^{*}\cancel{\text{atm}}} \times \frac{1^{*}\cancel{\text{N}\cdot\text{m}^{-2}}}{1^{*}\cancel{\text{Pa}}} \times \frac{(0.1^{*}\cancel{\text{m}})^{3}}{1^{*}\cancel{\text{ltr}}} \times \frac{1^{*}\text{J}}{1^{*}\cancel{\text{N}\cdot\text{m}}} = \underline{\underline{101.325 \text{ J}}}$$

$$R = 0.082054\,\cancel{\text{ltr}\cdot\text{atm}}\cdot\text{mol}^{-1}\cdot\text{K}^{-1} \times \frac{101.325\,\text{J}}{1^{*}\cancel{\text{ltr}\cdot\text{atm}}} = \underline{\underline{8.3141}}\ \frac{\text{J}\cdot\text{mol}^{-1}}{\text{K}}$$

$\underline{\underline{(1.9871 \text{ cal}\cdot\text{mol}^{-1}\cdot\text{K}^{-1})}}$

7.90

$$n = \frac{PV}{RT} = \frac{770/760\,\cancel{\text{atm}}\cdot 0.500^{*}\cancel{\text{ltr}}}{0.0821\,\cancel{\text{ltr}\cdot\text{atm}}\cdot\text{mol}^{-1}\cdot\cancel{\text{K}^{-1}}\cdot 273\cancel{\text{K}}} = \underline{0.0226 \text{ mol } O_2}$$

$\underline{\underline{2\ CO + O_2 \rightarrow 2\ CO_2}}$

(limiting reactant)

$$n = \frac{760/760\,\cancel{\text{atm}}\cdot 0.500^{*}\cancel{\text{ltr}}}{0.0821\,\cancel{\text{ltr}\cdot\text{atm}}\cdot\text{Mol}^{-1}\cdot\cancel{\text{K}^{-1}}\cdot 288\cancel{\text{K}}} = \underline{0.0211 \text{ mol CO}}$$

	(I)	(F)
P	760 torr	750 torr
V	500 ml	? ml
T	288 K	301 K

(Since mol CO consumed equal mol CO_2 produced the volume of the limiting reactant(CO) can be corrected to final conditions to obtain the final volume of CO_2.)

$$?\text{ml(F)} = 500^{*}\text{ml(I)}\cdot\frac{760\,\cancel{\text{torr}} \times 301\cancel{\text{K}}}{750\,\cancel{\text{torr}} \times 288\cancel{\text{K}}} = \underline{530\text{ml(F)}}$$

$\therefore$ $\underline{\underline{530\text{ml } CO_2}}$

7.91

$$?\text{mol } I_2 = 0.042\,\cancel{\text{ml soln}}\cdot\frac{0.0100\,\cancel{\text{mol } S_2O_3^{2-}}}{10^{3*}\cancel{\text{ml soln}}}\cdot\frac{1^{*}\text{mol } I_2}{2^{*}\cancel{\text{mol } S_2O_3^{2-}}} = \underline{\underline{2.1\cdot 10^{-7} \text{ mol } I_2}} \quad \text{(a)}$$

(b) $\underline{\underline{2.1\cdot 10^{-7}\text{mol } I_2 \text{ (SAME)}}}$

(c) $$?\text{mol } O_3 = 2.1\cdot 10^{-7}\,\cancel{\text{mol } I_2}\cdot\frac{1^{*}\text{mol } O_3}{1^{*}\cancel{\text{mol } I_2}} = \underline{\underline{2.1\cdot 10^{-7} \text{ mol } O_3}}$$

(d) $$?\text{ml(STP)} = 2.1\cdot 10^{-7}\,\cancel{\text{mol}}\cdot\frac{2.24\cdot 10^{4}\,\text{ml(STP)}}{1^{*}\cancel{\text{mol}}} = \underline{\underline{4.7\cdot 10^{-3}\text{ml(STP)}}}$$

(7.91 continued) (e)

$$?ppm\ O_3 = ?ltr\ O_3 = 10^6 * \cancel{ltr\ air} \cdot \frac{4.7 \times 10^{-6}\ ltr\ O_3}{2.0 \times 10^4\ \cancel{ltr\ air}} =$$

$\underline{2.4 \cdot 10^{-4} ppm\ O_3}$

7.92 The first step is to determine number of moles of NO by $n = \frac{PV}{RT}$

$$n = \frac{750/760\ \cancel{atm} \cdot 100 * \cancel{ltr}}{0.0821\ \cancel{ltr \cdot atm} \cdot mol^{-1} \cdot \cancel{K^{-1}} \cdot 773\cancel{K}} = 1.55\ \cancel{mol\ (NO)} \times \frac{5 * mol\ O_2}{4 * \cancel{mol\ NO}} = \frac{1.94 mol}{O_2}$$

The moles of O_2 required to produce 1.55 mol NO were calculated using the mole - mole relation shown by the balanced chemical equation. The next step is to use V = nRT/P to determine the volume of O_2.

$$V = 1.94\cancel{mol} \cdot 0.0821\ ltr \cdot \cancel{atm} \cdot \cancel{mol^{-1}} \cdot \cancel{K^{-1}} \cdot 298\cancel{K} / 0.895\ \cancel{atm} = \underline{53.1\ ltr (O_2)}$$

7.93 n = PV/RT

$$(a)\ n = \frac{(754-24)/760\ \cancel{atm} \cdot 0.0350\ \cancel{ltr}\ O_2}{0.0821\ \cancel{ltr \cdot atm} \cdot mol^{-1} \cdot \cancel{K^{-1}} \cdot 298\cancel{K}} = \frac{1.37 \cdot 10^{-3}}{mol\ O_2}$$

(b)

$$?g\ KClO_3 = 1.37 \cdot 10^{-3}\ \cancel{mol\ O_2} \cdot \frac{2 * \cancel{mol\ KClO_3}}{3 * \cancel{mol\ O_2}} \cdot \frac{122.6\ g\ KClO_3}{1 * \cancel{mol\ KClO_3}} = \underline{0.112 g\ KClO_3}$$

$$?\% = ?g\ KClO_3 = 100 * \cancel{g(s)} \cdot \frac{0.112 g\ KClO_3}{0.2500 \cancel{g(s)}} = \underline{44.8\ \%\ KClO_3}\quad (c)$$

7.94 (for one mole) $(P + \frac{a}{V^2})(V-b) = RT$ (R = 0.082054)

$$(V - b) = \frac{RTV^2}{PV^2 + a} \qquad V = \frac{RTV^2}{PV^2 + a} + b$$

The V^2 is part of a correction term for pressure. If we use the I.G. value for the molar volume (22.4 ltr) for the correction term we can calculate a more precise value for the volume than the I. G. value. If that more precise value is then used in the correction term we can calculate a still more precise value for V. By this method of successive approximations we can obtain as precise a value for V as the data and the van der Waals equation allow. (Usually only a few approximation steps are required.)

$$V_1 = \frac{0.082054\ \text{ltr}\cdot\cancel{\text{atm}}\cdot\text{mol}^{-1}\cdot\cancel{K^{-1}}\cdot 273.15\cancel{K}\cdot(22.41\cancel{\text{ltr}})^2}{1.0000\ \cancel{\text{atm}}\cdot(22.41\cancel{\text{ltr}})^2 + 1.36\ \cancel{\text{ltr}^2\cdot\text{atm}}} + 0.0318\ \text{ltr} = \underline{\underline{22.38\ \text{ltr/mol}}}$$

$$V_2 = \frac{0.082054\ \text{ltr}\cdot\cancel{\text{atm}}\cdot\text{mol}^{-1}\cdot\cancel{K^{-1}}\cdot 273.15\cancel{K}\cdot(22.38\ \cancel{\text{ltr}})^2}{1.0000\ \cancel{\text{atm}}\cdot(22.38\ \cancel{\text{ltr}})^2 + 1.36\ \cancel{\text{ltr}^2\cdot\text{atm}}} + 0.0318\ \text{ltr} = \underline{\underline{22.38}}$$

↑Since the precision of the value for 'a' limits the denominator to two decimal places, and thus four significant figures, the final answer is also limited to four S.F. In this example, to four S.F., the second approximation is identical to the first.

8.79 CH_4: $? \text{ kcal} = 1^* \text{deg} \times \frac{2.20^{\bullet} \text{ kcal}}{(273 - 161) \text{ deg}} = \underline{0.0196 \text{ kcal}}$

C_2H_6 : 0.0179 C_3H_8 : 0.0178 C_4H_{10} : 0.0195 C_6H_{14}: 0.0200

C_8H_{18} : 0.0204 $C_{10}H_{22}$: 0.0198   (all of these values round to 0.02)

ΔH_{vap} / B.P. $\approx$ CONSTANT FOR THIS SERIES

8.80 $? \text{ kcal} = 55.0^{\bullet} \text{ g (A)} \times \frac{9.22 \text{ kcal}}{1^* \text{mol (A)}} \times \frac{1 \text{ mol (A)}}{46.07 \text{ g(A)}} = \underline{11.0 \text{ kcal}}$

8.81 $? \text{ kJ} = 35.0^{\bullet} \text{ g Bz} \times \frac{2.37 \text{ kcal}}{1^* \text{mol Bz}} \times \frac{1^* \text{mol Bz}}{78.0 \text{ g Bz}} \times \frac{4.184^* \text{ kJ}}{1^* \text{kcal}} = \underline{4.45 \text{ kJ}}$

8.82 $? \text{ kJ} = 1 \text{ mol Hg} \times \frac{200.6 \text{ g Hg}}{1 \text{ mol Hg}} \times \frac{4.29 \text{ kJ}}{14.5 \text{ g Hg}} = \underline{59.3 \text{ kJ}}$ (14.2 kcal)

8.83

$$?\text{g ice} = 1/2^* \times 68.2\,\cancel{\text{kg}}(10^*\cancel{\text{mi}}\ \cancel{\text{hr}^{-1}})^2 \times \frac{1^*\cancel{\text{hr}^2}}{(60^*60^*\cancel{\text{s}})^2} \times \frac{(1609\ \cancel{\text{m}})^2}{(1^*\cancel{\text{mi}})^2} \times \frac{1\ \text{g ice}}{332\ \cancel{\text{J}}} = \underline{\underline{2.05\text{ g ice}}}$$

8.84

$$?\text{ cal} = 1.00\cancel{\text{g steam}} \times \frac{9710\ \text{cal}}{18.02\cancel{\text{g steam}}} + 60^*\cancel{\text{C}^\circ} \times 1.00\cancel{\text{g}} \times \frac{1^*\text{cal}}{1^*\cancel{\text{C}^\circ} \times 1^*\cancel{\text{g}}} =$$

(assumed 2 s.f.)

$\underline{\underline{599\text{ cal}}}$

8.85

HEAT LOST = HEAT GAINED

X = final temperature

$$\therefore\ 10.0\cancel{\text{g(Bz)}} \cdot \frac{2.37\cancel{\text{kcal}}}{78.0\cancel{\text{g(Bz)}}} = 50.0\cancel{\text{g } H_2O} \times \frac{10^{-3}\cancel{\text{kcal}}}{1^*\cancel{\text{g } H_2O}\cdot 1^*\cancel{\text{C}^\circ}} \cdot (30 - X)\cancel{\text{C}^\circ}$$

$\underline{\underline{X = 24^\circ\text{C}}}$

8.86

HEAT GAINED = HEAT LOST X = final temperature of water

$$50.0\cancel{\text{g ice}} \cdot \frac{1430\text{ cal}}{18.02\cancel{\text{g ice}}} + 50.0\cancel{\text{g } H_2O} \cdot X\cancel{\text{C}^\circ} \cdot \frac{1^*\text{cal}}{1^*\cancel{\text{g } H_2O}\cdot 1\cancel{\text{C}^\circ}} = 10.0\cancel{\text{g steam}} \times \frac{9710\text{ cal}}{18.02\cancel{\text{g steam}}} + 10.0\cancel{\text{g } H_2O} \cdot (100 - X)\cancel{\text{C}^\circ} \cdot \frac{1^*\text{cal}}{1^*\cancel{\text{g } H_2O}\cdot 1^*\cancel{\text{C}^\circ}}$$

$(3968 + 50.0X)\text{cal} = (5388 + 1000 - 10.0X)\text{cal}$ $60.0X = 2420$

$\underline{\underline{X = 40.3^\circ\text{C}}}$

8.87 (a) $n\lambda = 2\ d \sin\theta$ with $n = 1$ $\sin\theta = \lambda/2d = 0.1145$

angle = 6.6 degrees (b) angle = 34.9 degrees

8.88 $n\lambda = 2d \sin\theta$ $d = n\lambda/2 \sin\theta$ $?\text{pm} = \dfrac{1^* \bullet 0.141\ \text{nm}}{2^* \bullet \sin 20} = 206\ \text{pm}$

for 27.4 degrees: d = 153 pm for 35.8 degrees: d = 121 pm

8.89 $\theta = \sin^{-1}\left(\dfrac{n\lambda}{2\ d}\right) = \sin^{-1}\left(\dfrac{1^* \bullet 1.41\ \text{A}}{2^* \bullet 2.0\ \text{A}}\right) = 20.6$ degrees

(in phase for 2nd angle

for n = 2: 44.8 degrees for n = 3: (sin > 1) no reflection

(in phase for 3rd angle)

8.90

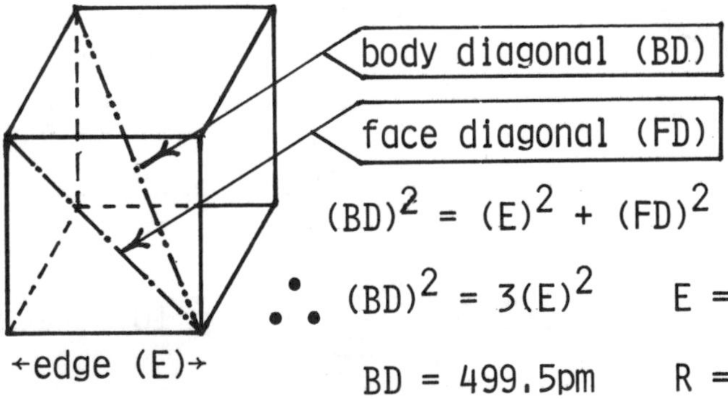

= 4 radii (R)

(for body-centered)

$(BD)^2 = (E)^2 + (FD)^2$ $(FD)^2 = (E)^2 + (E)^2$

$\therefore (BD)^2 = 3(E)^2$ E = 288.4pm $(BD)^2 = 3(288.4$

BD = 499.5pm R = BD/4 = 124.9pm

8.91 $?\text{Cr(at)} = 1^*\text{mol Cr} \bullet \dfrac{51.996\text{g Cr}}{1^*\text{mol Cr}} \bullet \dfrac{1^*\text{cm}^3\text{Cr}}{7.19\text{gCr}} \bullet \dfrac{1^*\text{unit cell}}{(2.884 \bullet 10^{-8}\text{cm})^3} \bullet$

$\dfrac{2\ \text{Cr(at)}}{1^*\text{unit cell}} = 6.03 \bullet 10^{23}\ \text{Cr(at)}$ atom = (at)

8.92 for face-centered cubic: face diagonal (FD) = 4 radii (R)

unit cell edge (E) = 407.86pm $\quad (FD)^2 = (E)^2 + (E)^2 = 2(E)^2 = (4R)^2$

$\therefore\ 4R = (E)\cdot 2^{1/2}$ $\quad$ (R) = 144.20pm

8.93 face diagonal (FD) = 4 radii (R) = 4 x 1.43Å = 5.72Å

edge (E) = $(FD)/2^{1/2}$ $\quad$ unit cell edge = 4.04Å

8.94 (refer to 8.90) $(BD)^2 = 3(E)^2 \quad (BD) = (E)\cdot 3^{1/2} = 412.3\cdot 1.732$pm

= 714.1pm $\quad (R)_{Cl^-} = 181$pm $\therefore\ (R)_{Cs^+} = (714.1 - 2\cdot(181))$pm/2 =

176pm

8.95 unit cell edge = 6.58Å $\quad (R)_{Cl^-} = 1.81$Å $\therefore\ (R)_{Rb^+} = 1.48$Å

(148pm)

8.96 for simple cubic; edge (E) = 2 radii (R)

for body-centered (see 8.90): $4(R) = (3(E)^2)^{1/2}$

for face-centered (see 8.92): $(FD) = 4(R) = (2(E)^2)^{1/2}$

(a) E = 288 pm $\quad$?g Ag = 1 ~~cm^3 Ag~~ $\cdot \dfrac{107.9\text{g Ag}}{6.022\cdot 10^{23}\ \text{Ag(at)}} \cdot \dfrac{1\ \text{Ag(at)}}{1\ \text{unit cell}} \cdot$

(THIS SET-UP IS CONTINUED ON THE NEXT PAGE.)

(8.96(a) continued) $\dfrac{1^{*}\text{unit cell}}{(288\text{pm})^3} \bullet \dfrac{(10^{10}\text{pm})^3}{(1^{*}\text{cm})^3} = \underline{\underline{7.50\text{g Ag}}}$ (density)

(b) $\underline{(E) = 333\text{pm}}$ density = ?g Ag = $1^{*}\text{cm}^3\text{Ag} \bullet \dfrac{107.9\text{g Ag}}{6.022 \bullet 10^{23}(\text{at})} \bullet \dfrac{2^{*}(\text{at})}{1^{*}\text{unitcell}}$

$\bullet \dfrac{1^{*}\text{unit cell}}{(333\text{pm})^3} \bullet \dfrac{(10^{10}\text{pm})^3}{(1^{*}\text{cm})^3} = \underline{\underline{9.70\text{g Ag}}}$ (density)

(c) $\underline{(E) = 407\text{pm}}$ density = ?g Ag = $1^{*}\text{cm}^3\text{Ag} \bullet \dfrac{107.9\text{g Ag}}{6.022 \bullet 10^{23}(\text{at})} \bullet \dfrac{4^{*}(\text{at})}{(407\text{pm})^3} \bullet$

$\dfrac{(10^{10}\text{pm})^3}{1^{*}\text{cm}^3} = \underline{\underline{10.6\text{g Ag}}}$ (density) $\therefore$ $\underline{\underline{\text{face-centered cubic structure for Ag}}}$

| 8.97 |

AY = BY = d sin θ   extra distance = $\underline{\underline{2d \sin\theta = n}}$
Bragg Equation

| 8.98 |

vacant space = volume of unit cell - volume of contained spher[e]

volume of a sphere = $4/3\ \pi r^3$ $\underline{r = 0.50\text{Å}}$ $\therefore$ $\underline{\text{volume} = 0.52\text{Å}^3}$

PRIMITIVE CUBIC: $\underline{\text{unit cell edge} = 1.0\text{Å}}$, $\underline{\text{unit cell volume} = 1.0\text{Å}^3}$

(unit cell contains one sphere) $\therefore$ vacant space = $(1.0 - 0.52)\text{Å}^3$ =
$\underline{\underline{0.48\text{Å}^3\ (48\%)}}$

BODY-CENTERED CUBIC: body diagonal = $4 \bullet 0.5\text{Å} = 2.0\text{Å}$ face diagonal =
$(2 \bullet \text{edge}^2)^{1/2}$ $(BD) = ((FD)^2 + (E)^2)^{1/2} = (3(E)^2)^{1/2} = (E) \bullet 3^{1/2} = 2.0\text{Å}$

(8.98 continued) $\therefore$ (E) = 2.0Å/$3^{1/2}$ = $\underline{1.15\text{Å}}$ volume = $(E)^3$ = $\underline{1.54\text{Å}^3}$

(unit cell contains two spheres) $\therefore$ vacant space = (1.54 - 2(0.52))Å^3 = $\underline{\underline{0.50\text{Å}^3}}$ (32%)

FACE-CENTERED CUBIC: (FD) = 4(r) = 2.0Å = (E)$\bullet 2^{1/2}$ (E) = $\underline{1.41\text{Å}}$

volume = $\underline{2.83\text{Å}^3}$ (unit cell contains four spheres) $\therefore$ vacant space =

(2.83 - 4(0.52))Å^3 = $\underline{\underline{0.75\text{Å}^3}}$ (26%)

8.99

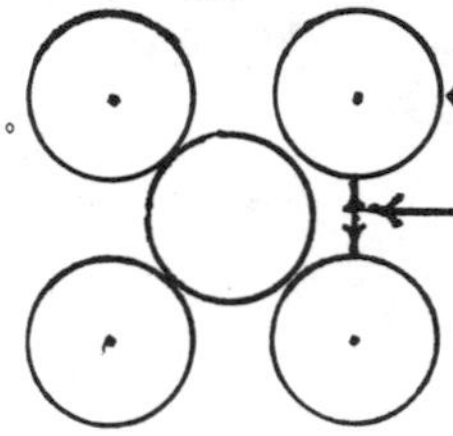

←Br^- (FD) = $4(R)_{Br^-}$ = (E)$\bullet 2^{1/2}$

←The Li^+ ion must fit here. $\therefore$ It must be this diameter (or smaller).

(E)$\bullet 2^{1/2}$ = 550pm $\bullet$ 1.414 = $\underline{777.8\text{pm}}$

$\therefore$ $\underline{\underline{(R)_{Br^-}}}$ = 777.8pm/4 = $\underline{\underline{194\text{pm}}}$ $\underline{\underline{(R)_{Li^+}}}$ =(550 - 2$\bullet$194)/2 = $\underline{\underline{81.0\text{pm}}}$ (or smaller)

8.100 face-centered cubic (see 8.98): volume = $(4.123\text{A})^3$ = $\underline{70.09\text{A}^3}$

(unit cell contains four Cs^+ and four Cl^-)

$$?gCsCl(FC) = 1^{*}\cancel{cm^3} \bullet \frac{168.36g\ CsCl}{6.0225\bullet 10^{23}\cancel{Cs^+Cl^-}} \bullet \frac{\cancel{4Cs^+Cl^-}}{70.09\cancel{A^3}} \bullet \frac{(10^8\cancel{A})^3}{1^{*}\cancel{cm^3}} = \underline{15.95g}$$

(≠ 3.99 so CsCl is not (FC))

body-centered cubic (see 8.98): volume = $\underline{70.09\text{A}^3}$

(unit cell contains 2 Cs^+ and 2 Cl^-) The density for this body-centered arrangement, calculated as shown above, equals $\underline{7.98}$ ≠ 3.99 (not (FC))

8.101 Since the volume of a unit cell equals the length of an edge cubed, the edge length can be calculated by taking the cube root of the unit cell volume:

$$?\text{nm} = 1^*\text{unit cell edge} = \left(\frac{1^*\not{cm}^3}{0.97\not{g\ Na}} \bullet \frac{22.99\not{g\ Na}}{6.022\bullet 10^{23}\not{Na(at)}} \bullet \frac{2\ \not{Na(at)}}{1^*\text{unit cell}} \bullet \frac{(10^7\text{nm})^3}{(1^*\not{cm})^3}\right)^{1/3} = \underline{\underline{0.43\text{nm}}}$$

8.102 $$?CaF_2(\text{units}) = 1^*\not{\text{unit cell}} \bullet \frac{(5.4626\not{Å})^3}{1^*\not{\text{unit cell}}} \bullet \frac{(1^*\not{cm})^3}{(10^8\not{Å})^3} \bullet \frac{3.180\not{g\,CaF_2}}{1^*\not{cm}^3} \bullet \frac{6.022\bullet 10^{23}\,CaF_2(\text{units})}{78.0768\not{g\ CaF_2}} = \underline{\underline{3.998\ CaF_2\ (\text{units})}}$$

8.103 The rock salt structure is a face-centered arrangement of Na^+ interpenetrated by a face-centered arrangement of Cl^-. The length of a unit cell edge is equal to the sum of 2 Na^+ radii plus 2 Cl^- radii. The edge length of a unit cell equals the cube root of the volume.

$$?\text{Å} = 1^*\text{unit cell edge} = \left(\frac{1^*\not{cm}^3}{2.165\not{g\ NaCl}} \bullet \frac{58.45\not{g\ NaCl}}{6.022\bullet 10^{23}\not{\text{ion pairs}}} \bullet \frac{4^*\not{\text{ion pairs}}}{1^*\text{unit cell}} \bullet \frac{(10^8\text{Å})^3}{1^*\not{cm}^3}\right)^{1/3} = \underline{5.64\text{Å}}$$

(If we assume that Na^+ and Cl^- touch along a unit cell edge, the radius of the Na^+ equals $(E)/2 - (R)_{Cl^-}$.)

$$(R)_{Na^+} = 5.64\text{Å}/2 - 1.81\text{Å} = \underline{\underline{1.01\ \text{Å}}}$$

PROBLEM SOLUTIONS FOR CHAPTER 10 OF BRADY & HUMISTON 3rd Ed.

10.29 $? X_{gly} = ?\ \text{mol gly} = 1^{*}\ \cancel{\text{mol sol}} \times \dfrac{45.0^{\bullet}\ \cancel{\text{g gly}}}{(45.0/92.0+100.0/18.02)\ \cancel{\text{mol sol}}}$

$\times \dfrac{1^{*}\ \text{mol gly}}{92.0\ \cancel{\text{g gly}}} = \underline{\underline{8.09 \times 10^{-2}\ \text{mol frctn glycine}}}$

(wt. frctn. = 0.310)

$?\ \text{wt \% gly} = ?\ \text{g gly} = 100^{*}\ \cancel{\text{g sol}} \times \dfrac{45.0^{\bullet}\ \text{g gly}}{(45.0 + 100^{*})\ \cancel{\text{g sol}}} = \underline{\underline{31.0\ \%\ \text{gly soltn}}}$

$?m = ?\text{mol gly} = \cancel{1^{*}\text{kg}\ H_2O} \times \dfrac{45.0^{\bullet}\ \cancel{\text{g gly}}}{100^{*}\ \cancel{\text{g}\ H_2O}} \times \dfrac{10^{3}\ \cancel{\text{g}\ H_2O}}{\cancel{1^{*}\text{kg}\ H_2O}} \times \dfrac{1^{*}\ \text{mol gly}}{92.0\ \cancel{\text{g gly}}} = \underline{\underline{4.89\ m}}$

10.30 (a) $?\%\ C_6H_6 = ?\ \text{g}\ C_6H_6 = 100^{*}\ \cancel{\text{g sol}} \times \dfrac{45.0\ \text{g}\ C_6H_6}{(45.0+80.0)\ \cancel{\text{g sol}}} = \underline{\underline{36.0\% C_6H_6}}$

(∴ 64.0%T)

(b)

$? X_{C_6H_6} = ?\text{mol}\ C_6H_6 = 1^{*}\ \cancel{\text{mol sol}} \times \dfrac{45.0^{\bullet}/78.11\ \text{mol}\ C_6H_6}{(45.0/78.11 + 80.0/92.14)\ \cancel{\text{mol sol}}} = \underline{\underline{0.399}}$

(mol frctn C_6H_6)

(∴ mol frctn Toluene = $\underline{\underline{0.601}}$)

(c)

$?\ m(Bz) = ?\ \text{mol Bz} = 10^{3*}\text{g Tol} \times \dfrac{45.0/78.11\ \text{mol Bz}}{80.0\ \text{g Tol}} = \underline{\underline{7.20\ m\ (Bz)}}$

10.31 (a)

$$? \text{ wt \%} = ? \text{ g } Zn(NO_3)_2 = 100^{*}\cancel{\text{g sol}} \times \frac{121.8^{\bullet}\text{g } Zn(NO_3)_2}{1000^{*}\cancel{\text{ml sol}}} \times \frac{1^{*}\cancel{\text{ml sol}}}{1.107\cancel{\text{g sol}}}$$

$= \underline{\underline{11.00\% \; Zn(NO_3)_2 \text{ solution}}}$

(b) $$? \text{ m} = ? \text{ mol } Zn(NO_3)_2 = \cancel{1^{*}\text{kg } H_2O} \times \frac{11.00^{\bullet}\text{g } \cancel{Zn(NO_3)_2}}{89.00\cancel{\text{g } H_2O}} \times \frac{10^{3}\cancel{^{*}\text{g } H_2O}}{\cancel{1^{*}\text{kg } H_2O}} \times$$

$$\frac{1 \text{ mol } Zn(NO_3)_2}{189.4\cancel{\text{g } Zn(NO_3)_2}} = \underline{\underline{0.6526 \text{ m } (Zn(NO_3)_2 \text{ solution}}}$$

(c) $$?X_{Zn(NO_3)_2} = ?\text{mol } Zn(NO_3)_2 = 1^{*}\cancel{\text{mol soln}} \bullet \frac{11.00^{*}/189.4 \text{ mol } Zn(NO_3)_2}{(11.00/189.4+89.00/18.02) \cancel{(\text{mol soln})}}$$

$= \underline{\underline{0.01162 \text{ (mole fraction)}}}$

(d) $$?M = ?\text{mol } Zn(NO_3)_2 = 1^{*}\cancel{\text{ltr soln}} \bullet \frac{121.8^{\bullet}\text{g } Zn(NO_3)_2}{1^{*}\cancel{\text{ltr soln}}} \bullet \frac{1^{*}\text{mol } Zn(NO_3)_2}{189.4\text{g } \cancel{Zn(NO_3)_2}}$$

$= \underline{\underline{0.6431 \text{ M}}}$

10.32 $$? X_{CuCl_2} = ?\text{mol } CuCl_2 = 1^{*}\cancel{\text{mol soln}} \bullet \frac{0.30^{\bullet}\text{mol } CuCl_2}{(0.30+40.0)\cancel{\text{mol soln}}} =$$

$\underline{\underline{7.4 \bullet 10^{-3} \text{(mole fraction)}}}$ (a)

(10.32 continued)(b) $?m = ?\text{mol } CuCl_2 = 1^{*}\text{kg } H_2O \bullet \dfrac{0.30^{\bullet}\text{mol } CuCl_2}{40.0\text{mol } H_2O} \bullet \dfrac{10^3/18.02 \text{ mol } H_2O}{1^{*}\text{kg } H_2O} = \underline{\underline{0.42 \text{ m } (CuCl_2)}}$

(c) $?\text{wt}\% = ?\text{g } CuCl_2 = 100^{*}\text{g soln} \bullet \dfrac{(0.30^{\bullet} \times 134.4)\text{g } CuCl_2}{(40.0 \bullet 18.01 + 0.30 \bullet 134.4)\text{g soln}} = \underline{\underline{5.3\% \ CuCl_2}}$

(DW) = drinking water

10.33 (a) $?\text{wt}\% = ?\text{g } CHCl_3 = 100^{*}\text{g(DW)} \bullet \dfrac{12.4^{\bullet}\text{g } CHCl_3}{10^6{}^{*}\text{g (DW)}} = \underline{\underline{1.24 \bullet 10^{-3}\%}}$

(assume that density (DW))

(b) $?M = ?\text{mol } CHCl_3 = 1^{*}\text{ltr(DW)} \bullet \dfrac{10^3{}^{*}\text{g(DW)}}{1^{*}\text{ltr(DW)}} \bullet \dfrac{12.4^{\bullet}\text{g } CHCl_3}{10^6{}^{*}\text{g (DW)}} \bullet \dfrac{1^{*}\text{mol} CHCl_3}{119.4\text{g} CHCl_3} = \underline{\underline{1.04 \bullet 10^{-4} \text{ M}}}$

10.34 $?\text{wt}\% = ?\text{g(IPA)} = 100^{*}\text{g soln} \bullet \dfrac{0.250^{\bullet} \times 60.10\text{g(IPA)}}{(0.250 \bullet 60.10 + 0.750 \bullet 18.02)\text{g soln}} = \underline{\underline{52.7\%\text{(IPA)}}}$

(b) $?m = ?\text{mol(IPA)} = 1^{*}\text{kg } H_2O \bullet \dfrac{0.250^{\bullet}\text{mol(IPA)}}{0.750 \text{ mol } H_2O} \bullet \dfrac{1^{*}\text{mol } H_2O}{18.02\text{g} H_2O} \bullet \dfrac{10^3{}^{*}\text{g } H_2O}{1^{*}\text{kg } H_2O} = \underline{18.5 \text{ m}}$

10.35 $?\underline{X} = ?\text{mol } NaHCO_3 = 1^{*}\text{mol soln} \bullet \dfrac{9.6^{\bullet}\text{g } NaHCO_3}{100^{*}\text{g } H_2O} \bullet \dfrac{1^{*}\text{mol } NaHCO_3}{84.01\text{g } NaHCO_3} \bullet$

(10.35 continued) $\dfrac{100^* \text{g } H_2O}{(100^*/18.02+9.6/84.02) \text{ mol soln}} = \underline{\underline{0.020 \text{ (mole frctn}}}$

$?m = ?\text{mol } NaHCO_3 = 1^* \text{kg } H_2O \bullet \dfrac{9.6^{\bullet} \text{g } NaHCO_3}{100^* \text{g } H_2O} \bullet \dfrac{1^* \text{mol } NaHCO_3}{84.01 \text{g } NaHCO_3} \bullet \dfrac{10^{3*} \text{g } H_2O}{1^* \text{kg } H_2O}$

$= \underline{\underline{1.1 \text{ m}}}$

10.36 $?\underline{X} = ?\text{mol NaCl} = 1^* \text{mol soln} \bullet \dfrac{6.25^{\bullet} \text{mol NaCl}}{1^* \text{kg } H_2O} \bullet \dfrac{1^* \text{kg } H_2O}{(10^{3*}/18.02+6.25 \text{ (mol soln}}$

$= \underline{\underline{0.101 \text{ (mole fraction)}}}$

$?\text{wt fraction} = ?\text{g NaCl} = 1^* \text{g soln} \bullet \dfrac{6.25^{\bullet} \text{mol NaCl}}{1^* \text{kg } H_2O} \bullet \dfrac{58.45 \text{g NaCl}}{1^* \text{mol NaCl}} \bullet$

$\dfrac{1^* \text{kg } H_2O}{(10^{3*}+6.25 \bullet 58.45) \text{g soln}} = \underline{\underline{0.268 \text{ (wt. fraction)}}}$

10.37 $?\underline{X} = ?\text{mol } Na_2CO_3 = 1^* \text{mol soln} \bullet \dfrac{14.0^{\bullet} \text{g } Na_2CO_3}{86.0 \text{g } H_2O} \bullet \dfrac{1^* \text{mol } Na_2CO_3}{106.0 \text{g } Na_2CO_3}$

$\dfrac{86.0 \text{g } H_2O}{(14.0/106.0+86.0/18.02) \text{mol soln}} = \underline{\underline{0.0269 \text{ (mole fraction)}}}$

$?m = ?\text{mol } Na_2CO_3 = 10^{3*} \text{g } H_2O \bullet \dfrac{14.0^{\bullet} \text{g } Na_2CO_3}{86.0 \text{g } H_2O} \bullet \dfrac{1^* \text{mol } Na_2CO_3}{106.0 \text{g } Na_2CO_3} = \underline{\underline{1.54 \text{ m}}}$

10.38 (a) $?M = ?\text{mol Mg} = 1^{*}\cancel{\text{ltr(DW)}} \cdot \dfrac{0.150^{\bullet}\cancel{\text{g Mg}}}{1^{*}\cancel{\text{ltr(DW)}}} \cdot \dfrac{1^{*}\text{mol Mg}}{24.31\cancel{\text{g Mg}}} = \underline{\underline{6.17 \cdot 10^{-3}\ M}}$

b) $?m = ?\text{mol Mg} = 1^{*}\cancel{\text{kg } H_2O} \cdot \dfrac{1^{*}\cancel{\text{ltr } H_2O}}{1^{*}\cancel{\text{kg } H_2O}} \cdot \dfrac{0.150^{\bullet}\cancel{\text{g Mg}}}{1^{*}\cancel{\text{ltr } H_2O}} \cdot \dfrac{1^{*}\text{mol Mg}}{24.31\cancel{\text{g Mg}}} =$

$\underline{\underline{.17 \cdot 10^{-3} m}}$ NOTE: For very dilute solutions molarity and molality are essentially equal.

10.39 (a) $?m = ?\text{mol } H_2SO_4 = 10^{3*}\cancel{\text{g } H_2O} \cdot \dfrac{96.0\cancel{\text{g } H_2SO_4}}{4.0^{\bullet}\cancel{\text{g } H_2O}} \cdot \dfrac{1^{*}\text{mol } H_2SO_4}{98.08\cancel{\text{g } H_2SO_4}}$

$= \underline{\underline{240\ m}}$ ←←←NOTE - only 2 s.f.

(b) $?\underline{X} = ?\text{mol } H_2SO_4 = 1^{*}\cancel{\text{mol soln}} \cdot \dfrac{(96.0/98.08)\text{mol } H_2SO_4}{(96.0/98.08+4.0/18.02)^{\bullet}\cancel{\text{mol soln}}} =$

$= \underline{\underline{0.815 \text{ (mole fraction)}}}$ (c) $\therefore \underline{X}_{(H_2O)} = 1^{*} - 0.815^{\bullet} = \underline{\underline{0.185 \text{ (mol frctn)}}}$

10.40 (a) $?\text{wt\%} = ?\text{g } NH_4NO_3 = 100^{*}\cancel{\text{g soln}} \cdot \dfrac{2.25^{\bullet}\cancel{\text{mol } NH_4NO_3}}{10^{3*}\cancel{\text{g } H_2O}} \cdot$

$\dfrac{30.04\text{g } NH_4NO_3}{1^{*}\cancel{\text{mol } NH_4NO_3}} \cdot \dfrac{10^{3*}\cancel{\text{g } H_2O}}{(10^{3}+2.25 \cdot 80.04)\cancel{\text{g soln}}} = \underline{\underline{15.3\%\ NH_4NO_3}}$

(b) $?\underline{X} = ?\text{mol } NH_4NO_3 = 1^{*}\cancel{\text{mol soln}} \cdot \dfrac{2.25^{\bullet}\text{mol } NH_4NO_3}{(2.25+10^{3}/18.02)\cancel{\text{mol soln}}} = \underline{\underline{0.0390 \text{ (mol frctn)}}}$

(c) $\therefore \underline{X}_{(H_2O)} = 1^{*} - 0.0390^{\bullet} = \underline{\underline{0.9610 \text{ (mole fraction)}}}$

10.41 $\underline{X}_{(CHCl_3)} = 1^* - \underline{X}_{(C_6H_6)} = 1^* - 0.240 = \underline{0.760 \text{ (mole fraction)}}$

(a) $?\text{mol\%} = ?\text{mol CHCl}_3 = 100^*\text{mol soln} \cdot \frac{0.760\text{ mol CHCl}_3}{1^*\text{mol soln}} = \underline{\underline{76.0\text{mol\%}}}$

(b) $?m = ?\text{mol } C_6H_6 = 10^{3*}\text{g CHCl}_3 \cdot \frac{1^*\text{mol CHCl}_3}{119.4\text{g CHCl}_3} \cdot \frac{0.240\text{ mol } C_6H_6}{0.760\text{mol CHCl}_3} =$

$\underline{\underline{2.65\text{ m}}}\ (C_6H_6)$

(c) $?m = ?\text{mol CHCl}_3 = 10^{3*}\text{g } C_6H_6 \cdot \frac{1^*\text{mol } C_6H_6}{78.11\text{g } C_6H_6} \cdot \frac{0.760\text{mol CHCl}_3}{0.240\text{mol } C_6H_6} = \underline{\underline{40.5\text{m}}}\ (CHCl_3)$

(d) $?\text{wt\%} = ?\text{g CHCl}_3 = 100^*\text{g soln} \cdot \frac{0.760\text{mol CHCl}_3}{1^*\text{mol soln}} \cdot \frac{119.4\text{g CHCl}_3}{1^*\text{mol CHCl}_3} \cdot$

$\frac{1^*\text{mol soln}}{(0.760 \cdot 119.4 + 0.240 \cdot 78.11)\text{ g soln}} = \underline{\underline{82.9\%\ CHCl_3}} \quad \therefore \quad \underline{\underline{17.1\%\ C_6H_6}}$

10.42 (EG) = ethylene glycol (the 200g H_2O is assumed to be exact)*

$?m = ?\text{mol(EG)} = 1^*\text{kg } H_2O = \frac{222.6\text{ g(EG)}}{200^*\text{g } H_2O} \cdot \frac{1^*\text{mol(EG)}}{62.07\text{g(EG)}} \cdot \frac{10^{3*}\text{g } H_2O}{1^*\text{kg } H_2O} = \underline{\underline{17.93\text{m}}}$

$?M = ?\text{mol(EG)} = 1^*\text{ltr soln} \cdot \frac{222.6\text{ g(EG)}}{(222.6+200^*)\text{g soln}} \cdot \frac{1^*\text{mol(EG)}}{62.07\text{g(EG)}} \cdot \frac{1072\text{g soln}}{1^*\text{ltr soln}}$

$= \underline{\underline{9.097\text{ M}}}$

10.43 $?wt\% = ?g(EG) = 100^{*}\cancel{g\ soln} \bullet \frac{4.03^{\bullet}\cancel{mol(EG)}}{10^3{}^{*}\cancel{ml\ soln}} \bullet \frac{62.07\,g(EG)}{1^{*}\cancel{mol\ (EG)}} \bullet \frac{1^{*}\cancel{ml\ soln}}{1.045\cancel{g\ soln}} = \underline{\underline{23.9\%\ (EG)}}$

$?\underline{X}_{(EG)} = ?mol(EG) = 1^{*}\cancel{mol\ soln} \bullet \frac{4.03\,mol(EG)}{1045\cancel{g\ soln}} \bullet \frac{1045\cancel{g\ soln}}{(4.03+0.761 \bullet 1045/18.02)\cancel{mol\ soln}} = \underline{\underline{0.0836\ (\text{mole fraction (EG)})}}$

$?m = ?mol(EG) = 1^{*}\cancel{kg\ H_2O} \bullet \frac{\cancel{4.03}^{\bullet}mol(EG)}{(1045-4.03 \bullet 62.07)\cancel{g\ H_2O}} \bullet \frac{10^3{}^{*}\cancel{g\ H_2O}}{1^{*}\cancel{kg\ H_2O}} = \underline{\underline{5.07\ m\ (EG)}}$

10.44 $?kJ = 10.0^{\bullet}\cancel{g\ AlCl_3} \bullet \frac{1^{*}\cancel{mol\ AlCl_3}}{133.3\cancel{g\ AlCl_3}} \bullet \frac{321\,kJ}{1^{*}\cancel{mol\ AlCl_3}} = \underline{\underline{24.1\ kJ}}$

10.45 $?cal = 115\cancel{g\ NH_4NO_3} \bullet \frac{26^{\bullet}\cancel{kJ}}{1^{*}\cancel{mol\ NH_4NO_3}} \bullet \frac{1^{*}\cancel{mol\ NH_4NO_3}}{80.04\cancel{g\ NH_4NO_3}} \bullet \frac{10^3{}^{*}cal}{4.184\cancel{kJ}} = \underline{\underline{8900\ cal}}$

10.46 @ 70°C saturated $NaNO_3$ solution = 137g/100*g H_2O, @ 25°C = 93g

IN 200*g H_2O: 274g $NaNO_3$ →→→→ 186 g $NaNO_3$ ∴ <u>88g $NaNO_3$ precipitates</u>

10.47 $?g\ H_2O = 35.0^{\bullet}\cancel{g\ NaBr} \bullet \frac{100^{*}g\ H_2O}{118\cancel{g\ NaBr}} = \underline{\underline{30.g\ H_2O\ (80^{\circ}C)}}$

10.48 Eth = ethane

$$?P_{Eth} = ?\text{ torr Eth} = 5.00^{\bullet} \times 10^{-2} \cancel{\text{g Eth}} \bullet \frac{751 \text{ torrEt}}{6.56 \bullet 10^{-2} \cancel{(\text{g Eth})}}$$

$= \underline{572 \text{ torr Eth}}$

10.49 $?\text{torr(gas)} = 0.0478 \cancel{\text{g} \bullet \text{ltr}^{-1}} \bullet (6.50 \bullet 10^{-5} \cancel{\text{g} \bullet \text{ltr}^{-1}} \bullet \text{torr}^{-1})^{-1} =$

$\underline{735 \text{ torr(gas)}}$ P_{H_2O} @ 25 C = $\underline{24 \text{ torr}}$ $\therefore P_T = (735+24) \text{torr} = \underline{759\text{to}}$

10.50 (GW) = ground water

$$? \text{ g } CH_4 \bullet \text{ltr}^{-1} = 10^{3} \cancel{{}^{*}\text{atm}} \bullet \frac{2.09^{\bullet} \times 10^{-4} \text{gCH}}{0.968 \cancel{\text{atm}}}$$

$= \underline{0.216\text{g } CH_4 \text{ (per ltr (GW))}}$

10.51 For a solution which follows Raoult's law, the vapor pressure of the solvent in equilibrium with the solution is equal to th vapor pressure of the solvent at that temperature times the mole fraction of the solvent in the solution.

$$?\text{torr} = 93.4^{\bullet} \text{torr} \bullet \frac{1000^{*}/78.11 \text{ mol } C_6H_6}{(1000^{*}/78.11 + 56.4/282.6)\text{mol soln}} = \underline{92.0 \text{ torr}}$$

↑↑↑↑↑↑↑↑↑↑↑↑↑↑↑↑↑↑↑↑↑↑↑↑↑↑↑
the mole fraction of benzene in the solution

10.52 The mole fraction of methyl alcohol in the solution is equal to the vapor pressure of methyl alcohol over the solution divided by the vapor pressure of pure methyl alcohol at the same temp.

$X_{meth} = 130 \cancel{\text{torr}}/160 \cancel{\text{torr}} = \underline{0.813}$ $\therefore \underline{X_{gly} = 0.187}$

10.53 $P_{soln} = X_{H_2O} \bullet P^{o}_{H_2O}$ (Raoult's law) $\therefore P^{o}_{H_2O} = X^{-1}_{H_2O} \bullet P_{soln}$

$$?X_{H_2O} = \frac{(100^*/18.02)\ \text{mol}\ H_2O}{(100^*/18.02 + 150/92.10)\ \text{mol soln}} = \underline{0.773\ (\text{mole fraction}\ H_2O)}$$

$$P^{o}_{H_2O} = \frac{91.1\ \text{torr}}{0.773} = \underline{117.8\ \text{torr}}$$

From table 7.1; $\underline{P^{o}_{H_2O}\ @55^{\circ}C = 118\text{tor}}$

$\underline{\underline{55^{\circ}C}}$

10.54 $P_T = P_{hep} + P_{oct}$ $P_{hep} = X_{hep} \bullet P^{o}_{hep}$ $P_{oct} = X_{oct} \bullet P^{o}_{oct}$

$$?\text{mol hep} = 25.0^{\bullet}\ \text{g hep} \bullet \frac{1^*\text{mol hep}}{100.2\ \text{g hep}} = \underline{0.2495\ \text{mol hep}}$$

$$?\text{mol oct} = 35.0^{\bullet}\ \text{g oct} \bullet \frac{1^*\text{mol oct}}{114.2\ \text{g oct}} = \underline{0.3064\ \text{mol oct}}$$

$$P_T = \frac{0.2495}{(0.2495 + 0.3064)} \bullet 791\ \text{torr} + \frac{0.3064}{(0.2495 + 0.3064)} \bullet 352\ \text{torr} = \underline{\underline{549\text{torr}}}$$

10.55
$$?\ C = -1.86C^{\circ} \bullet \frac{1^*\text{kg}\ H_2O}{1^*\text{mol Glyc}} \bullet \frac{55.0^{\bullet}\ \text{g Glyc}}{250\ \text{g}\ H_2O} \bullet \frac{1^*\text{mol Glyc}}{92.10\text{g Glyc}} \bullet \frac{10^{3*}\text{g}H_2O}{1^*\text{kg}\ H_2O}$$

$= \underline{-\ 4.44C^{\circ}\ (\text{f.p. lowering})}$ $\therefore$ f.p. $= 0^{\circ}C - 4.44C^{\circ} = \underline{\underline{-4.44^{\circ}C}}$

$\therefore$ boiling point $= \underline{\underline{101.2^{\circ}C}}$

10.56 $?MW = ?gX = 1^{*}\text{mol X} \cdot \frac{1^{*}\text{kg } C_6H_6 \cdot 5.12 C^{\circ}}{1^{*}\text{mol X}} \cdot \frac{3.84 \text{ g X}}{0.500 \text{ kg } C_6H_6 \cdot 0.307 C^{\circ}}$

= 128g X ←←←approximate MW of X   The empirical formula wt. = 64.07

∴ MW = 128.1   molecular formula = $C_8H_4N_2$

(S) =substance

10.57 $? MW = ?g\ (S) = 1^{*}\text{mol(S)} \cdot \frac{1^{*}\text{kg } H_2O \cdot 1.86 C^{\circ}}{1^{*}\text{mol(S)}} \cdot \frac{16.9 \text{ g(S)}}{250^{*}\text{g } H_2O} \cdot \frac{10^3{*}\text{g } H_2O}{1^{*}\text{kg } H_2O}$

$\cdot \frac{1^{*}}{0.744 C^{\circ}} = 169g\ (S)$ (The molecular weight of the substance.)

CALCULATE THE EMPIRICAL FORMULA: $?\text{mol C} = 57.2 \text{ g C} \cdot \frac{1^{*}\text{mol C}}{12.01 \text{ g C}} = 4.76 \text{ mol C}$

$?\text{mol H} = 4.77 \text{ g H} \cdot \frac{1^{*}\text{mol H}}{1.008 \text{ g H}} = 4.73 \text{ mol H}$ (2.38 mol O)

∴ C_2H_2O ←←←empirical formula→→→ 42.04g ∴ molecular formula = $C_8H_8O_4$

(G) = glucose

10.58 $?g(G) = 0.750 C^{\circ} \cdot 150^{*}\text{g } H_2O \cdot \frac{1^{*}\text{mol(G)}}{1^{*}\text{kg } H_2O \cdot 1.86 C^{\circ}} \cdot \frac{180.2 \text{ g(G)}}{1^{*}\text{mol(G)}} \cdot$

$\frac{1^{*}\text{kg } H_2O}{10^3{*}\text{g } H_2O} = 10.9g(G)$ $?C^{\circ}(\Delta b.p.) = -0.750 C^{\circ}(\Delta f.p.) \cdot \frac{0.51 C^{\circ}(\Delta b.p.)}{-1.86 C^{\circ}(\Delta f.p.)}$

$= 0.21\ C^{\circ}(\Delta b.p.)$ ∴ boiling point = 100.21°C

10.59 $\Delta(B.P.) = 0.51\cdot \cancel{C^\circ} \cdot \frac{2.47\ C^\circ}{1.86\ \cancel{C^\circ}} = \underline{0.68\ C^\circ}$ $\therefore$ $\underline{\underline{B.P. = 100.68^\circ C}}$

10.60 $\Delta(f.p.) = \frac{-1.86\cdot C^\circ}{1^*\cancel{m\ soln}} \cdot 0.1075\ \cancel{m\ soln} = \underline{-.0200\ C^\circ}$ $\underline{\underline{f.p. = -0.200^\circ C}}$

10.61 $\Pi = MRT$ $?atm = 0.0821\ \cancel{ltr}\cdot atm\cdot \cancel{mol^{-1}}\cdot \cancel{K^{-1}}\cdot 298\cancel{K}\cdot \frac{5.0\cdot \cancel{g(S)}}{1^*\cancel{ltr}}\cdot \frac{1^*\cancel{mol(S)}}{342\cancel{g(S)}}$

$= \underline{\underline{0.36\ atm}} \cdot \frac{760\ torr}{1^*atm} = \underline{\underline{272\ torr}}$

10.62 $MW = grams\ /\ mol$ $\Pi V = nRT$ n = number of moles

$\therefore$ $MW = \frac{grams}{n} = \frac{grams\ \cdot\ RT}{\Pi V} = \frac{0.40\cdot g\cdot 0.0821\ \cancel{ltr}\cdot \cancel{atm}\cdot 300\cancel{K}}{3.74/760\ \cancel{atm}\cdot 1.00\ \cancel{ltr}\cdot mol\cdot \cancel{K}} = \underline{\underline{2.0\cdot 10^3 g\ mol^{-1}}}$

10.63 $\Pi = 0.020\cdot \cancel{mol}\cdot \cancel{ltr^{-1}}\cdot 0.0821\ \cancel{ltr}\cdot \cancel{atm}\cdot \cancel{mol^{-1}}\cdot \cancel{K^{-1}}\cdot 298\cancel{K}\cdot \frac{760\ torr}{1^*\cancel{atm}} =$

↑ (effective molarity for 100% dissociation)

$\underline{\underline{370\ torr}}$ ←←(only 2 s.f.)

10.64 $i_{HF} = 1.91\ \cancel{C^\circ}/1.86\ \cancel{C^\circ} = 1.03$ ('i' factors for weak electrolytes are only slightly larger than 1.)

10.65 $?g\ Cl^- = 3.78\ \text{ltr(SW)} \bullet \dfrac{0.566\ \text{mol}\ Cl^-}{1000^*g\ H_2O} \bullet \dfrac{35.45g\ Cl^-}{1^*\text{mol}\ Cl^-} \bullet \dfrac{1000^*g\ H_2O}{(10^{3*}+36.35)\ (g(SW))}$

$\bullet \dfrac{1024g(SW)}{1^*1\text{ltr(SW)}} = \underline{\underline{74.9g\ Cl^-}}$

(Associated with 1000g(SW) are the moles of species as shown in the table. See calculation below.)

$(0.566\bullet35.45+0.486\bullet22.99+0.055\bullet24.31+0.029\bullet96.06+0.011\bullet40.08+0.011\bullet39.10$
$+0.002\bullet61.02)g = \underline{36.35g}$ (Total mass of all dissolved ions in 1000g(SW).)

$(41.7g\ Na^+)$ $(4.99g\ Mg^{2+})$ $(10.4g\ SO_4^{2-})$ $(1.7g\ Ca^{2+})$ $(1.6g\ K^+)$ $(0.5g\ HCO_3^-)$

TOTAL MASS OF IONS = 60.9

10.66 $?M\ Cl^- = ?\ \text{mol}\ Cl^- = 1^*1\text{ltr soln} \bullet \dfrac{1024g\ \text{soln}}{1^*1\text{ltr soln}} \bullet \dfrac{0.566\text{mol}\ Cl^-}{(10^{3*}+0.566\bullet35.5)\ (g\ \text{soln})}$

$= \underline{0.568\ M\ (Cl^-)}$

NOTE -(The calculated molarity is close to the molality. For Na^+ the calculated molarity is 0.492M. For the ions present in more dilute condition, molarities may be assumed equal to molalities.)

$M_{total} = \underline{1.168}$

$\Pi = RTM$

$?\Pi = ?atm = 1.168\text{mol}\bullet\text{ltr}^{-1}\bullet0.0821\ \text{ltr}\ atm\ \text{mol}^{-1}\bullet K^{-1}\bullet298^{\bullet}K = \underline{\underline{28.6\ atm}}$

$\therefore$ The minimum pressure for reverse osmosis would have to exceed 28.6.

10.67 If the solute were completely dissociated the effective molality would be equal to 0.20.

$?\Delta(f.p.) = ?\ C° = 0.20^{\bullet}\text{mol}\bullet(\text{kg}\ H_2O)^{-1}\bullet \dfrac{1^*\text{kg}\ H_2O \bullet 1.86\ C°}{1^*\text{mol}} = \underline{0.37\ C°}$

(10.67 continued) Assuming complete dissociation the expected freezing point would be -0.37°C.

Table 10.6 gives an 'i' factor of 1.21 for 0.1 m $MgSO_4$.

THE LOWERING OF THE FREEZING POINT CALCULATED USING THIS 'i' FACTOR IS EQUAL TO: $1.21 \cdot 0.10^{\bullet} \times (-1.86\ C^{o}) = \underline{-0.23\ C^{o}}$ f.p. = $\underline{\underline{-0.23^{o}C}}$

10.68 (H) = hydrate

$$?g(H) = 50.0^{\bullet}\cancel{g\ soln} \cdot \frac{10.0\cancel{g\ NaCO_3}}{100^{*}\cancel{g\ soln}} \cdot \frac{(106+180)g(H)}{106\cancel{g\ NaCO_3}}$$

$= \underline{\underline{13.5g(H)}}$

10.69

$$?g\ O_2 = 1^{*}\cancel{ltr\ soln} \cdot \frac{5.34 \cdot 10^{-5}g\ O_2}{1^{*}\cancel{ltr\ soln} \cdot 1^{*}\cancel{torr\ O_2}} \cdot 760^{*}\cancel{torr(air)} \cdot \frac{20^{\bullet}\cancel{torr\ O_2}}{100^{*}\cancel{torr(air)}} = \underline{\underline{8.1 \cdot 10^{-3}g\ O_2}}$$

10.70

$$P_{total}(25\ C) = 93.4^{\bullet}torr \cdot \frac{60.0/78.1\ mol\ C_6H_6}{(60.0/78.1+40.0/92.1)mol\ soln} + 26.9^{\bullet}torr \cdot \frac{40.0/92.1\ mol\ C_7H_8}{(60.0/78.1+40.0/92.1)mol\ soln} = \underline{\underline{69.4\ torr}}$$

10.71 Starting with the desired final answer, the molecular weight (MW) of the unknown substance (UN), one could employ a reasoned series of steps as follows: Since the mass of the (UN) (43.3g) is given the (MW) could be calculated if the number of moles of the (UN) could be determined. The number of moles of (UN) could be determined if the mole fraction of (UN), the mole fraction of CCl_4 and the number of moles of

(10.71 continued) CCl_4 can be found. The mole fractions can be calculated by using Raoult's law and the given vapor pressure data. (As you can see from the problem solution below it takes as much space to explain the problem as it does to solve it!)

If the mole fraction (UN) = X, then the mole fraction CCl_4 = 1-X.

$$P_{CCl_4} = X_{CCl_4} \bullet P^{o}_{CCl_4}$$

$$P_{(UN)} = X_{(UN)} \bullet P^{o}_{(UN)} \qquad P_{total} = P_{CCl_4} + P_{(UN)}$$

$$137 \text{ torr} = (1-X) \bullet 143 \text{ torr} + X \bullet 85 \text{ torr} \qquad (137-143) = (85-143) \bullet X$$

X = 0.103 (mole fraction (UN)) If 43.3g(UN) represent a mole fraction of 0.103 and a mole fraction of 0.897(CCl_4) corresponds to the number of moles represented by (400*/153.8)(CCl_4) then:

$$?(MW)(UN) = ?g(UN) = 1^{*}\cancel{\text{mol(UN)}} \bullet \frac{43.3^{\bullet}\text{g(UN)}}{(400^{*}/153.8)\cancel{\text{mol } CCl_4}} \bullet \frac{0.897\cancel{\text{mol } CCl_4}}{0.103\cancel{\text{mol (UN)}}} =$$

150 (MW)(UN)

10.72 $P_{CHCl_3} = P^{o}_{CHCl_3} \bullet X_{CHCl_3} \quad X_{CHCl_3} = P_{CHCl_3}/P^{o}_{CHCl_3} = 511^{\bullet}/526$

= 0.971 (mole fraction $CHCl_3$) $\therefore$ $X_{(S)}$ = 0.029 (mole fraction (S))

(SEE 10.71)

(b) $$?\text{mol(S)} = 8.3^{\bullet}\cancel{\text{g(S)}} \bullet \frac{0.029\text{mol(S)}}{0.971\cancel{\text{mol } CHCl_3}} \bullet \frac{1^{*}\cancel{\text{mol } CHCl_3}}{8.3\cancel{\text{g (S)}}} = 0.030\text{mol(S)}$$

$$?(MW) = ?g(S) = 1^{*}\cancel{\text{mol(S)}} \bullet \frac{8.3^{\bullet}\text{g(S)}}{0.030\cancel{\text{mol(S)}}} = 280 \text{ g(S) (molecular weight (S))}$$

10.73 1 ltr of H_2O = 1kg H_2O 1 ltr (EG) = 1113g (EG)

$$?\Delta(\text{f.p.}) = ?\ C^\circ = -1.86\ C^\circ \cdot \frac{1\ \text{kg } H_2O}{1\ \text{mol (EG)}} \cdot \frac{1113\ \text{g (EG)}}{1\ \text{kg } H_2O} \cdot \frac{1\ \text{mol (EG)}}{62.07\ \text{g (EG)}} =$$

$-33.4\ C^\circ$ ∴ f.p. = $-33.4^\circ C$ ($-28^\circ F$) Although the first ice crystals might form at $-28^\circ F$ the solution becomes more concentrated as water is removed by freezing and the F.P. would drop to a lower temperature so that the engine might indeed be safe down to a temperature of $-34^\circ F$ or below.

10.74 Let $m^\bullet$ = the particle molality. $m^\bullet = (1+\alpha)m$ (α=fraction dissociated)

$$\alpha = \frac{m^\bullet}{m} - 1$$

$$?m^\bullet = -0.500\ C^\circ \cdot \frac{1\ m^\bullet}{-1.86\ C^\circ} = 0.2688 \qquad \alpha = 0.2688/0.250 - 1 = 1.08 - 1$$

= 0.08 (fraction dissociated) ∴ 8% dissociated

PROBLEM SOLUTIONS FOR CHAPTER 11 of BRADY & HUMISTON 3rd Ed.

11.38 $V_2 = ?,\ V_3 = ?$ (T & N = constants)

$? \text{ l}(2) = 10.0^{\bullet}\text{l (1)} \times \frac{15.0}{7.50} = \underline{20.0 \text{ l (2)}}$ $? \text{ l}(3) = 10.0 \text{ l}(1) \times \frac{15.0}{1.00} =$

$\underline{150 \text{ l (3)}}$ $w_{1-2} = P_r \Delta V = 7.50^{\bullet}\text{atm} \times 10.0\text{ l} = \underline{75.0 \text{ l}\bullet\text{atm}}$

$w_{2-3} = 1.00^{\bullet}\text{atm} \times 130 \text{ l} = \underline{130 \text{ l}\bullet\text{atm}}$ $\Delta T = 0 \ \therefore\ \Delta E = 0 \ \therefore\ q = w$ (ideal gas)

$\therefore\ q_{1-2} = \underline{75.0 \text{ l}\bullet\text{atm}}$ $q_{2-3} = \underline{130 \text{ l}\bullet\text{atm}}$ $\Delta E_{surr} = 0$

11.39 $w = P_r \Delta V$ $1^*\text{kJ} = 1^*\text{kPa}\bullet\text{m}^3$ $V_2 = P_1V_1/P_2$

$\therefore\ V_2 = 50.0^{\bullet}\text{m}^3 \times \frac{200 \text{ kPa}}{100 \text{ kPa}} = \underline{100 \text{ m}^3}$ $?\text{kJ} = 100 \text{ kPa} \times (100-50.0)^{\bullet}\text{m}^3 =$ (2 s.f.)

(2 s.f.)
$\underline{5000 \text{ kJ}}$ for Ideal Gas if $\Delta T = 0 \ \therefore\ \Delta E = 0 \ \therefore\ \underline{q = w = 5000 \text{ kJ}}$

11.40 $w = P_{ext}\Delta V = 3.00^{\bullet}\text{atm} \times (250^*-500^*)\text{ml} \times \frac{2.42 \times 10^{-5}\text{ kcal}}{1^*\text{ml}\bullet\text{atm}} =$

(11.40 continued) $\underline{-0.0182\ kcal}$ $\underline{q = 3.00\ kcal}$ $\Delta E = q - w =$

$(3.00^{\bullet} + 0.0182)(kcal) = \underline{3.02\ kcal}$ $\underline{\underline{\Delta E_{surr} = -\ \Delta E_{syst} = -3.02\ kcal}}$

11.41 $\Delta E = -3.41$ kcal $?°C = 25.000(C) + 3.41^{\bullet} \times 10^{3} \cancel{cal} \times \dfrac{1^{*}(C°)}{4\ 250\ \cancel{cal}} =$

$\underline{\underline{25.802°C}}$

11.42 (a) $?cal = (27.282 - 25.000)\cancel{C°} \bullet \dfrac{23.2^{\bullet} \times 10^{3}(cal)}{1\cancel{C°}} = \underline{\underline{5.29 \bullet 10^{4} cal}}$

$\Delta E = ?kcal = 1^{*}\cancel{mol\ C_3H_8} \bullet \dfrac{-52.9^{\bullet}(kcal)}{0.100\cancel{mol\ C_3H_8}} = \underline{\underline{-529 kcal}}$

11.43 $\Delta H = \Delta E + \Delta(PV)$ $\Delta(PV) = (\Delta n)RT$ $\Delta n = 3mol\ Prod_{(g)} - 6mol$

$React_{(g)} = -3mol$ $\therefore$ $\Delta H = \{ -52.9^{\bullet}\cancel{kcal} + (-3^{*}\cancel{mol}) \bullet 1.987 \bullet 10^{-3} \cancel{kcal}(\cancel{mol} \bullet \cancel{K})^{-1}$

$\bullet 298\cancel{K}\} \bullet \dfrac{4.184(kJ)}{1^{*}\cancel{kcal}} = \underline{\underline{-2210 kJ}}$ (3 s.f.)

11.44 $\Delta H = q_p$ $\Delta E = \Delta H - \Delta(PV)$ $\Delta(PV) = \Delta(nRT)$

$q_p = \frac{-38.6kcal}{0.5*mol\ OF_2} = \underline{\underline{-77.2kcal = \Delta H}}$ since T = constant (25°C) $\therefore$

$\Delta(nRT) = \Delta n(RT)$ Δn = +1mol(gas) (from the equation) $\therefore$ $\Delta n(RT) =$

$1*\cancel{mol}\bullet 1.987\bullet 10^{-3}(kcal)(\cancel{mol}\bullet\cancel{K})^{-1}\bullet 298\cancel{K} = \underline{\underline{0.592kcal}}$ $\Delta E = -77.2kcal -$

$0.592kcal = \underline{\underline{-77.8kcal}}$

11.45 (1*atm) (298K) $\underline{q_p = -15.6kcal} = \Delta H$ $\Delta E = \Delta H - \Delta(PV)$

@ constant P: $\Delta(PV) = P\Delta V$ $\Delta V = V_f - V_i$ $\Delta V = 74.10\cancel{g}\bullet \frac{1*ml}{2.24\cancel{g}} -$

$56.08\cancel{g}\bullet \frac{1*ml}{3.25\cancel{g}} - 18.02\cancel{g}\bullet \frac{1*ml}{0.997\cancel{g}} = \underline{-2.25ml}$ $\therefore$ $P\Delta V = \underline{-2.25ml\bullet atm} =$

$\underline{-5.45\bullet 10^{-5}kcal}$ Since $\Delta H = -15.6kcal\bullet mol^{-1}$, the $P\Delta V$ term which is the difference between ΔH and ΔE is much smaller than the uncertainty ($\pm$ $1\bullet 10^{-1}$) associated with the ΔH value! $\therefore$ $\Delta H \simeq \Delta E$ when all of the substances involved are liquids and/or solids.

11.46 $H_{2(g)} + 1/2\ O_{2(g)} \longrightarrow H_2O_{(l)}$ $\Delta E_f^\circ = \Delta H_f^\circ - \Delta n(RT)$ (see 11.44)

(11.46 continued) $\Delta n = -1\ 1/2{*}mol$ $\Delta E_f^\circ = -286\,kJ - (-1.5{*}\cancel{mol} \cdot 8.315 \cdot 10^{-3}\,kJ \cdot \cancel{mol}^{-1} \cdot \cancel{K}^{-1} \cdot 298\cancel{K}) = \underline{\underline{-282kJ}}$

$?calorimeter\ constant = ?KJ = 1{*}\cancel{C^\circ} \cdot \frac{0.200\cancel{mol\ H_2O_{(l)}} \cdot (-282\,kJ \cdot \cancel{mol}^{-1})}{0.880\ \cancel{C^\circ}} = \underline{\underline{-64.1kJ}}$

$\Delta E = ?kJ = 1{*}\cancel{mol\ C_7H_8} \cdot \frac{0.615{*}\cancel{C^\circ}}{0.0100\cancel{mol\ C_7H_8}} \cdot \frac{-64.1\,kJ}{1{*}\cancel{C^\circ}} = \underline{\underline{-3940kJ}}$

11.47 $2\ C_{(s)} + 1/2\ O_{2(g)} + 3\ H_{2(g)} \longrightarrow C_2H_5OH_{(l)}$ $\Delta H_f^\circ\ (C_2H_5OH_{(l)})$

$\cancel{2\ CO_{2(g)}} + \cancel{3\ H_2O_{(l)}} \longrightarrow C_2H_5OH_{(l)} + \cancel{3\ O_{2(g)}}$ $\Delta H = 1.37 \cdot 10^3 kJ$

$2\ C_{(s)} + \cancel{2\ O_{2(g)}} \longrightarrow \cancel{2\ CO_{2(g)}}$ $2 \cdot \Delta H_f^\circ = -788kJ$

$3\ H_{2(g)} + \cancel{3/2}\ O_{2(g)}$ (1/2) $\longrightarrow \cancel{3\ H_2O_{(l)}}$ $3 \cdot \Delta H_f^\circ = -858kJ$

$\Delta H_f^\circ(C_2H_5OH_{(l)}) = \underline{\underline{-276kJ}}$

11.48 $3\ C_{(s)} + 4\ H_{2(g)} \longrightarrow C_3H_{8(g)}$ $\Delta H_f^\circ = ?$

$\cancel{3\ CO_{2(g)}} + \cancel{4\ H_2O_{(l)}} \longrightarrow C_3H_{8(g)} + \cancel{5\ O_{2(g)}}$ (11.43) $\Delta H_f^\circ = 2220kJ$

(11.48 continued) $3\ C_{(s)} + 3\ O_{2(g)} \longrightarrow 3\ CO_{2(g)}$ $\quad 3 \bullet \Delta H_f^\circ = -1182\text{kJ}$

$4\ H_{2(g)} + 2\ O_{2(g)} \longrightarrow 4\ H_2O_{(l)}$ $\quad 4 \bullet \Delta H_f^\circ = -1144\text{kJ}$

$\Delta H_f^\circ(C_3H_{8(g)}) = -106\text{kJ}$

11.49 $7\ C_{(s)} + 4\ H_{2(g)} \longrightarrow C_7H_{8(l)}$ $\quad \Delta H_f^\circ = ?$

$7\ CO_{2(g)} + 4\ H_2O_{(l)} \longrightarrow C_7H_{8(l)} + 9\ O_{2(g)}$ $\quad$ (see 11.46) $-\Delta H_{comb.} = 3940\text{kJ}$

$7\ C_{(s)} + 7\ O_{2(g)} \longrightarrow 7\ CO_{2(g)}$ $\quad 7 \bullet \Delta H_f^\circ = -2758\text{kJ}$

$4\ H_{2(g)} + 2\ O_{2(g)} \longrightarrow 4\ H_2O_{(l)}$ $\quad 4 \bullet \Delta H_f^\circ = -1144\text{kJ}$

$\Delta H_f^\circ(C_7H_{8(l)}) = 38\text{kJ}$

11.50 $\Delta E = \Delta H - \Delta(PV)$ $\quad T = 298\text{K}$ $\quad \Delta(PV) = \Delta n(RT)$ $\quad \Delta n = -2\text{mol}$

$$\therefore\ \Delta E = -214.3\text{kcal} - \left(-2\text{*mol} \bullet \frac{1.987 \bullet 10^{-3}\text{kcal} \bullet 298\text{K}}{1\text{*mol} \bullet \text{K}}\right) = -213.1\text{kcal}$$

11.51 $\Delta H_{vap}(H_2O)$ @ 25 C = $10.5 kcal \cdot mol^{-1}$ $\Delta(PV) = P(V_{vap} - V_{liq}) \approx$ PV_{vap} (assume ideal gas) $PV_{vap} = RT = 1.987 \cdot 10^{-3} kcal \cdot mol^{-1} \cdot K^{-1} \cdot 298K =$ $0.592 kcal \cdot mol^{-1}$ $\Delta E_{vap} = \Delta H_{vap} - \Delta(PV) = (10.5 - 0.592) kcal \cdot mol^{-1} =$ $9.9 kcal \cdot mol^{-1}$ $q_p = \Delta H = 10.5 kcal \cdot mol^{-1}$ $w = P\Delta V = 0.592 kcal \cdot mol^{-1}$

11.52 $2\ Al_{(s)} + Fe_2O_{3(s)} \longrightarrow Al_2O_{3(s)} + 2\ Fe_{(s)}$ $\Delta H = ?$

$$2\ Al_{(s)} + \cancel{1\ 1/2\ O_{2(g)}} \longrightarrow Al_2O_{3(s)} \qquad \Delta H_f = -1676 kJ$$

$$Fe_2O_{3(s)} \longrightarrow \cancel{1\ 1/2\ O_{2(g)}} + 2\ Fe_{(s)} \qquad -\Delta H_f = 822.2 kJ$$

(a) $\Delta H = -854 kJ$

(b) $-1429 kJ$ (c) $-402 kJ$ (d) $-87 kJ$ (e) $-136 kJ$

11.53 (a) $\Delta H = \Delta H_f(C_2H_4) - \Delta H_f(C_2H_2) - \Delta H_f(H_2) = (12.4 - 54.2 - 0)$ $kcal \cdot mol^{-1} = -41.8 kcal \cdot mol^{-1}$ (b) $-31.6 kcal \cdot mol^{-1}$ (c) $-40.8 kcal \cdot mol^{-1}$ (d) $9.9 kcal \cdot mol^{-1}$ (e) $-684 kcal \cdot mol^{-1}$

11.54 The ΔH° for the net reaction is equal to the sum of the ΔH° for step (1) and the ΔH° for step (2). $\Delta H^\circ = (-30 - 64)$ kcal

11.55 $2\,N_{2(g)} + 5\,O_{2(g)} \longrightarrow 2\,N_2O_{5(g)}$ $\quad \Delta H_{rctn} = ?$

$2 \times (2\,HNO_{3(l)} \longrightarrow N_2O_{5(g)} + H_2O_{(l)})$ $\quad 2\bullet\Delta H = 36.6$ kcal

$4\bullet(1/2\,N_{2(g)} + 3/2\,O_{2(g)} + 1/2\,H_{2(g)} \longrightarrow HNO_{3(l)})$ [5/4 written above 3/2] $\quad 4\bullet\Delta H = -166$ kcal

$2\,H_2O_{(l)} \longrightarrow 2\,H_{2(g)} + O_{2(g)}$ $\quad -\Delta H = 136.6$ kcal

$\Delta H_{rctn} = 7$ kcal

11.56 $FeO_{(s)} + CO_{(g)} \longrightarrow Fe_{(s)} + CO_{2(g)}$ $\quad \Delta H = ?$

$1/3\bullet(3FeO_{(s)} + CO_{2(g)} \longrightarrow Fe_3O_{4(s)} + CO_{(g)})$ $\quad 1/3\bullet\Delta H = -12.7$ kJ

$1/6\bullet(2Fe_3O_{4(s)} + CO_{2(g)} \longrightarrow 3Fe_2O_{3(s)} + CO_{(g)})$ $\quad 1/6\bullet\Delta H = 9.8$ kJ

$1/2\bullet(Fe_2O_{3(s)} + 3CO_{(g)} \longrightarrow 2Fe_{(s)} + 3CO_{2(g)})$ [3 corrected to 2] $\quad 1/2\bullet\Delta H = -14.0$ kJ

$\Delta H = -17$ kJ

11.57 $Fe_{(s)} + 1/2\ O_{2(g)} \longrightarrow FeO_{(s)}$ $\qquad \Delta H_f = ?$

(from 11.56) $\bigcirc\!\!\!Fe_{(s)} + \cancel{CO_{2(g)}} \longrightarrow \bigcirc\!\!\!FeO_{(s)} + \cancel{CO_{(g)}}$ $\qquad \Delta H = 4.00 kcal/mol$

$\cancel{C_{(s)}} + \tfrac{1}{2}\bigcirc\!\!\!O_{2(g)} \longrightarrow \cancel{CO_{2(g)}}$ $\qquad \Delta H = -94.1 kcal/mol$

$\cancel{CO_{(g)}} \longrightarrow \cancel{C_{(s)}} + \cancel{1/2\ O_{2(g)}}$ $\qquad \Delta H = 26.4 kcal/mol$

$\underline{(267 kJ)}$ $\qquad\qquad \underline{\Delta H_f(FeO_{(s)}) = -63.7 kcal\cdot}$
$\qquad\qquad\qquad\qquad\qquad\qquad\qquad \underline{mol^{-1}}$

11.58 $2C_{(graph)} + H_{2(g)} \longrightarrow C_2H_{2(g)}$ $\qquad \Delta H_f = ? kcal\cdot mol^{-1}$

$\cancel{CaC_{2(s)}} + \cancel{2H_2O_{(l)}} \longrightarrow \bigcirc\!\!\!C_2H_{2(g)} + \cancel{Ca(OH)_{2(s)}}$ $\quad \Delta H = -30.0$

$\cancel{Ca(OH)_{2(s)}} + \longrightarrow \cancel{CaO_{(s)}} + H_2O_{(l)}$ $\qquad \Delta H = 15.6$

$\cancel{CaO_{(s)}} + \textcircled{2}\bigcirc\!\!\!C_{(s)} \longrightarrow \cancel{CaC_{2(s)}} + \cancel{CO_{(g)}}$ $\qquad \Delta H = 110.5$

$1/2\cdot\cancel{(2CO_{(g)}} \longrightarrow \cancel{2C_{(s)}} + \cancel{O_{2(g)}}$ $\qquad 1/2\cdot\Delta H = 26.4$

$1/2\cdot\cancel{(O_{2(g)}} + \bigcirc\!\!\!2H_{2(g)} \longrightarrow \cancel{2H_2O_{(l)}}$ $\qquad 1/2\cdot\Delta H = -68.3$

$\qquad\qquad \underline{\Delta H_f = 54.2 kcal\cdot mol^{-1}}$ $\qquad \underline{(227 kJ\cdot mol^{-1})}$

11.59 $CaSO_4 \bullet 1/2\ H_2O_{(s)} + 3/2\ H_2O_{(l)} \longrightarrow CaSO_4 \bullet 2H_2O_{(s)}$ $\Delta H^\circ = ?$

$\Delta H^\circ = \Delta H_f^\circ(CaSO_4 \bullet 2H_2O_{(s)}) - \Delta H_f^\circ(CaSO_4 \bullet 1/2\ H_2O_{(s)}) - 3/2 \bullet \Delta H_f^\circ(H_2O_{(l)}) =$

$-2020 kJ \bullet mol^{-1} + 1573 kJ \bullet mol^{-1} - 3/2 \bullet (-286 kJ \bullet mol^{-1}) = \underline{\underline{-18.0 kJ \bullet mol^{-1}}}$

11.60 (a) $2NO_{(g)} + O_{2(g)} \longrightarrow 2NO_{2(g)}$ $\Delta H^\circ = [2(34)-0-2(90.4)]\ kJ$

$= \underline{\underline{-113 kJ}}$ (b) $\underline{\underline{305 kJ}}$ (c) $\underline{\underline{-106 kJ}}$

11.61 (1) $C_2H_5OH + 1/2\ O_2 \longrightarrow CH_3CHO + H_2O$ $\Delta H^\circ = -167-286+278+0 = \underline{\underline{-175 kJ}}$ (2) $CH_3CHO + 1/2\ O_2 \longrightarrow CH_3COOH$ $\Delta H^\circ = -487+167+0 = \underline{\underline{-320 kJ}}$ (3) $CH_3COOH + 2O_2 \longrightarrow 2CO_2 + 2H_2O$ $\Delta H^\circ = 2(-394)+2(-286) + 487 = \underline{\underline{-873 kJ}}$ NET EQUATION: $C_2H_5OH + 3O_2 \longrightarrow 2CO_2 + 3H_2O$

$\Delta H^\circ_{net} = \Delta H_1^\circ + \Delta H_2^\circ + \Delta H_3^\circ = -175 - 320 - 873 = \underline{\underline{-1368 kJ}}$

11.62 $C_2H_{6(g)} + 7/2\ O_{2(g)} \longrightarrow 2CO_{2(g)} + 3H_2O_{(g)}$ $\Delta H^\circ = ?$

$\Delta H^\circ = 2(-94.1)+3(-57.8)-(-20.2)\ kcal \bullet mol^{-1} = \underline{-341.4 kcal \bullet mol^{-1}}$

(11.62 continued) $?\text{cal} = 45.0\text{g } C_2H_{6(g)} \bullet \frac{341.4 \bullet 10^3 \text{ cal}}{30.07\text{g } C_2H_{6(g)}} = \underline{5.11 \bullet 10^5 \text{cal}}$

11.63 $?\text{g carbohydrate} = 2000\text{kcal} \bullet \frac{1\text{g carbohydrate}}{4\text{kcal}} = \underline{500\text{g carbohydrate}}$

(only 1 s.f.)

11.64 $H_2O_{(l)} \longrightarrow H_2O_{(g)}$ $\Delta H^\circ_{vap} = (-242 + 286)\text{kJ} = \underline{44\text{kJ}}$ (table 11.11

$? \text{kJ} = 10.0\text{ g } H_2O \bullet \frac{44 \text{ kJ}}{18.02\text{g } H_2O} = \underline{24\text{kJ}}$

11.65 $C_8H_{18(l)} + 12.5O_{2(g)} \longrightarrow 8CO_{2(g)} + 9H_2O_{(l)}$ $\Delta H^\circ = ?$

$C_8H_{18(l)} \longrightarrow 8C_{(s)} + 9H_{2(g)}$ $-\Delta H^\circ_f = 208.4\text{kJ}$

$8 \bullet (C_{(s)} + O_{2(g)} \longrightarrow CO_{2(g)}$ $8 \bullet \Delta H^\circ_f = -3150 \text{ kJ}$

$9 \bullet (H_{2(g)} + 1/2\ O_{2(g)} \longrightarrow H_2O_{(l)}$ $9 \bullet \Delta H^\circ_f = -2570 \text{ kJ}$

$\Delta H^\circ = -5510\text{kJ} \bullet \text{mol}^{-1}$

$?\text{kJ} = 3.79\text{ ltr } C_8H_{18} \bullet \frac{703\text{g } C_8H_{18}}{1^*\text{ ltr } C_8H_{18}} \bullet \frac{-5510\text{ kJ}}{114.2\text{g } C_8H_{18}} = \underline{-1.29 \bullet 10^5\text{kJ}}$

(11.65 continued) $?g\ H_2 = -1.29 \cdot 10^5 \cancel{kJ} \cdot \frac{2.016g\ H_2}{-286\ \cancel{kJ}} =$ $\underline{909g\ H_2}$

$?ltr\ H_2(F) = 909\cancel{g\ H_2} \cdot \frac{22.4\cancel{ltr\ H_2(STP)}}{2.016\cancel{g\ H_2}} \cdot \frac{1^*\cancel{atm} \cdot 298\cancel{K} \cdot ltr\ H_2(F)}{170\cancel{atm} \cdot 273\cancel{K} \cdot \cancel{1ltr\ H_2(STP)}} =$

$\underline{64.9\ ltr\ H_2(F)}$ $\underline{17.1gal\ H_2}$ A car would require a fuel tank 17x larger and able to contain 200 atm pressure! (This doesn't sound practical!)

11.66 $?g\ H_2O(evap) = 5900^*\cancel{kJ} \cdot \frac{18.02g \cdot H_2O(evap)}{(-242+286)\ \cancel{kJ}} =$ $\underline{2400g\ H_2O\ (evap)}$

(ONLY 2 s.f.)

11.67 $?g\ glucose = 5900\cancel{kJ(XS)} \cdot \frac{100^*\cancel{kJ(total)}}{60^*\cancel{kJ(XS)}} \cdot \frac{180.2g\ glucose}{2820\ \cancel{kJ(total)}} =$

$\underline{630g\ glucose}$

11.68 $\Delta H^\circ(\text{combustion of } CH_4) = \Delta H_f^\circ(CO_{2(g)}) + 2 \cdot \Delta H_f^\circ(H_2O_{(g)}) - \Delta H_f^\circ(CH_{4(g)})$

$\Delta H^\circ = (-94.1 - 2 \cdot 57.8 + 17.9)kcal \cdot mol^{-1} = \underline{-191.8kcal \cdot mol^{-1}}$

$?\ ltr\ CH_4(F) = 250^*\cancel{ml\ H_2O} \cdot 1.00\cancel{g\ H_2O_{(l)}}/1^*\cancel{ml\ H_2O_{(l)}}\ \cancel{(20^\circ C)}$ • (continued)

(11.68 continued) $\left(\frac{80.0\text{ cal}}{1\text{ g } H_2O(20^\circ C \rightarrow 100^\circ C)} + \frac{539.6\text{ cal}}{1\text{ g } H_2O(\text{evap})}\right) \cdot \frac{1\text{ mol } CH_{4(g)}}{191.8 \cdot 10^3\text{ cal}}$

$$\frac{22.4\text{ ltr } CH_4(STP)}{1\text{ mol } CH_{4(g)}} \cdot \frac{298K \quad 1\text{ tr}(F)}{273K \quad 1\text{ tr}(STP)} = \underline{\underline{19.8\text{ ltr } CH_{4(g)}(F)}}$$

11.69 $NaCl_{(s)} \longrightarrow Na^+_{(g)} + Cl^-_{(g)} \quad \Delta H^\circ = ?\text{kJ}$

(ionization energy) $Na_{(g)} \longrightarrow Na^+_{(g)} + 1e^- \quad \Delta H^\circ = 494.1\text{kJ}$

(electron affinity) $Cl_{(g)} + 1e^- \longrightarrow Cl^-_{(g)} \quad \Delta H^\circ = 351\text{ kJ}$

(sublimation energy) $Na_{(s)} \longrightarrow Na_{(g)} \quad \Delta H^\circ = 108\text{ kJ}$

(dissociation energy) $1/2\, Cl_{2(g)} \longrightarrow Cl_{(g)} \quad 1/2\Delta H^\circ = 121\text{ kJ}$

$NaCl_{(s)} \longrightarrow Na_{(s)} + 1/2Cl_{2(g)} \quad -\Delta H_f^\circ = 413\text{ kJ}$

(LATTICE ENERGY OF NaCl) $\quad \Delta H^\circ = 785\text{ kJ}$

11.70 $\Delta H^\circ = \Delta H_f^\circ(NO_{(g)}) + \Delta H_f^\circ(O_{(g)}) - \Delta H_f^\circ(NO_{2(g)}) = (90.4+249-34)\frac{\text{kJ}}{\text{mol}^{-1}}$

$= \underline{305\text{kJ}} \qquad \Delta E^\circ = \Delta H^\circ - \Delta(PV) \qquad \Delta(PV) = \Delta n(RT) \qquad \Delta E^\circ = \Delta H^\circ + \Delta n(RT)$

$\Delta n = +1 \qquad \Delta E^\circ = 305\text{kJ}\cdot\text{mol}^{-1} - 1\cdot 8.314\cdot 10^{-3}\text{kJ}\cdot\text{mol}^{-1}\cdot K^{-1}\cdot 298K = \underline{303\text{kJ/mol}}$

(11.70 continued) $?\text{kJ}\bullet\text{molecule}^{-1} = 303\text{kJ}\bullet\text{mol}^{-1}\bullet\dfrac{1*\text{mol}}{6.022\bullet10^{23}\text{molecule}}\bullet\dfrac{10^3*\text{J}}{1*\text{kJ}} = \underline{\underline{5.03\bullet10^{-19}\text{J}\bullet\text{molecule}^{-1}}}$ $\Delta E = h\nu$ $c = \nu\lambda$ $\Delta E = hc/\lambda$ $\therefore$

$\lambda = hc/\Delta E$ $\lambda = 6.626\bullet10^{-34}\text{J}\bullet\text{s}\bullet2.998\bullet10^{8}\text{m}\bullet\text{s}^{-1} / 5.03\bullet10^{-19}\text{J} =$

$\underline{\underline{3.95\bullet10^{-7}\text{m}}}$ $\underline{\underline{(395\ \text{nm})}}$

11.71 $H_{2(g)} + 2C_{(s)} \longrightarrow C_2H_{2(g)}$ $\Delta H_f^\circ = ?\text{kJ}$

$-\Delta H_f^\circ(C_2H_2)$ = E(bonds formed) - E(bonds broken) = $2\bullet$E(H-C) + E(C≡C) - E(H-H) -2E(atomization $C_{(s)}$) = $2\bullet415$kJ + 833kJ - 436kJ - $2\bullet715$kJ

$\underline{\underline{\Delta H_f^\circ = 203\text{kJ}}}$

11.72 ΔH_f°(calculated) = $6\bullet$E(H-C) + $3\bullet$E(C-C) + $3\bullet$E(C=C) - $3\bullet$E(H-H) - $6\bullet$E(atomization $C_{(s)}$) =$-6\bullet415$kJ $-3\bullet348$kJ + $3\bullet607$kJ + $3\bullet436$kJ + $6\bullet715$kJ = $\underline{243\ \text{kJ}}$ Since the determined $\Delta H_f^\circ(C_6H_6)$= $\underline{82.8\text{kJ}\bullet\text{mol}^{-1}}$, the difference is $\underline{\underline{160\ \text{kJ}}}$ (RESONANCE ENERGY). Species like benzene are much more stable than predicted due to the so called "resonance energy".

11.73 $3H_{2(g)} + 3C_{(s)} \longrightarrow C_3H_{6(g)}$ $\quad\Delta H_f^\circ = ?\text{kJ}$

$3\bullet(H_{2(g)} \longrightarrow$ ~~$2H_{(g)}$~~$)$ $\quad 6\bullet\Delta H_f^\circ = 1308\ \text{kJ}$

$3\bullet(C_{(s)} \longrightarrow$ ~~$C_{(g)}$~~$)$ $\quad 3\bullet\Delta H^\circ = 2145\ \text{kJ}$

~~$6H_{(g)}$~~ $+$ ~~$3C_{(g)}$~~ $\longrightarrow C_3H_{6(g)}$ $\quad {}^*\Delta H_T = -3445\ \text{kJ}$

$^*(\Delta H_T = 6\bullet E(C\text{-}H) + E(C{=}C) + E(C\text{-}C)) = -2490 - 607 - 348 = -3445\ \text{kJ})$

$\Delta H_f^\circ = 8\ \text{kJ}$

11.74 ΔH_f°(calculated) = ?kJ bonds formed: 8(C-H), 2(C-C)

In the graphite structure each carbon atom is bonded to other carbons by two single and one double bond and can thus be assigned one half of that total amount of energy. There are three moles of carbon in one mole of C_3H_8. ∴ bonds broken: $3\bullet(1/2)\bullet(2\bullet E(C\text{-}C) + E(C{=}C))$, $4\bullet E(H\text{-}H)$

$\Delta H_f^\circ(\text{calculated}) = 8\bullet(-415) + 2\bullet(-348) - 3\bullet(1/2)(2\bullet(-348) + (-607)) -$

$4\bullet(-436) = \underline{\underline{-317.5\ \text{kJ}}}$ The value from table 11.1 is -104 kJ which is only 1/3 of the calculated value but our description of the graphite bonding did not include the resonance energy associated with that structure. This problem illustrates the use of bond energies to calculate heats of formation can give very inaccurate results when resonance is involved.

11.75 $\Delta S = \frac{q_{rev}}{T}$ (since these processes are at constant pressure $q_P = q_{rev} = \Delta H \therefore \Delta S = \Delta H/T$)

$\Delta S_{vap} = 9720 cal \bullet mol^{-1}/373K = \underline{26.1 cal \bullet mol^{-1} \bullet K^{-1}}$ $\Delta S_{fus} = \frac{1440 cal \bullet mol^{-1}}{273°K} =$ $\underline{5.27 cal \bullet mol^{-1} \bullet K^{-1}}$ Both melting and vaporization involve increases in randomness ($\Delta S = +$), but vaporization also produces a very large increase in volume with an accompanying additional increase in entropy.

11.76 $\Delta G = \Delta H - \Delta(TS)$ FOR CONSTANT T: $\Delta G = \Delta H - T\Delta S$ AT EQUILIBRIUM: $\Delta G = 0 \therefore \Delta H = T\Delta S$ $\Delta H_{vap}(Br_2) = \underline{7.39 kcal \bullet mol^{-1}}$ $\Delta S_{vap}(Br_2) = (58.65 - 36.38) cal \bullet K^{-1} = \underline{22.27 cal \bullet K^{-1}}$ $\therefore 7.39 \bullet 10^3 cal \bullet mol^{-1} = T \bullet 22.27 cal/mol \bullet K$

$\underline{T = 331.8K}$ $\underline{(58.8^oC)}$

11.77 (a) There is a net decrease of 0.5 moles of gas for both reactions and the reactions are similar so a difference in the reaction entropies can not be predicted without calculation.

$\Delta S_{(a)} = 61.3 cal(mol \bullet K)^{-1} - (59.3 + (1/2)(49.0)) cal(mol \bullet K)^{-1} = \underline{-22.5 \frac{cal}{mol \bullet K}}$

$\Delta S_{(b)} = (51.06 - (47.3 + (1/2)(49.0))) cal/mol \bullet K = \underline{-20.7 cal \bullet mol^{-1} \bullet K^{-1}}$

$\therefore$ Reaction (a) is accompanied by the greater entropy change.

11.78 (a) $\Delta G^\circ = \Delta G_f^\circ(Al_2O_{3(s)}) + 2\bullet\Delta G_f^\circ(Fe_{(s)}) - 2\bullet\Delta G_f^\circ(Al_{(s)}) - \Delta G_f^\circ(Fe_2O_{3(s)})$

$\Delta G^\circ = -1577kJ + 2\bullet(0) - 2\bullet(0) - (-741.0kJ) = \underline{\underline{-836kJ}}$ (b) $\underline{\underline{-346kJ}}$ (c) $\underline{\underline{-101kJ}}$

11.79 $\Delta G^\circ = \Delta H^\circ - T\Delta S^\circ \therefore \Delta S^\circ = (\Delta H^\circ - \Delta G^\circ)/T$ $\quad \underline{T = 298K}$

(a) $\Delta S^\circ = (-854kJ - (-836kJ))\bullet 10^3 J/1 kJ / 298K = \underline{\underline{-60J/K}}$

(b) $\underline{\underline{\Delta S^\circ = -188J/K}}$ (c) $\underline{\underline{\Delta S^\circ = 47J/K}}$

11.80 $\Delta G = 6\bullet\Delta G_f(CO_{2(g)}) + 6\bullet\Delta G_f(H_2O_{(l)}) - \Delta G_f(C_6H_{12}O_{6(s)}) - \Delta G_f(O_2)$

$\Delta G = (6\bullet(-94.3) + 6\bullet(-56.7) - (-217.54) - (0))kcal\bullet mol^{-1} = \underline{\underline{-688.5 \frac{kcal}{mol}}}$

11.81 maximum useful work $= \Delta G = (3\bullet(-395)+4\bullet(1228)-(-23)-5\bullet(0))kJ =$

$\underline{\underline{-2074kJ}}$ The maximum useful work is obtained under reversible conditions. Any real process can only approach reversibility so the actual work for any real process will always be less than ΔG.

11.82 (a) $\Delta G^\circ = 12.4kcal$ (NO) (b) $\Delta G^\circ = -13.8kcal$ (YES) (c) $\Delta G^\circ = -122.7kcal$ (YES) (d) $\Delta G^\circ = 22.1kcal$ (NO) (e) $\Delta G^\circ = -8.3kcal$ (YES) (f) $\Delta G^\circ = 393kcal$ (NO)

PROBLEM SOLUTIONS FOR CHAPTER 12 of BRADY & HUMISTON 3rd Ed.

12.42 $CH_{4(g)} + 2O_{2(g)} = CO_{2(g)} + 2H_2O_{(g)}$ $\quad \frac{-\Delta[CH_4]}{\Delta t} = 0.16 \cdot \frac{mol}{(l \cdot s)}$

$$\frac{\Delta[CO_2]}{\Delta t} = \frac{-\Delta[CH_4]}{\Delta t} = \underline{\underline{0.16 \text{ mol } (l \cdot s)^{-1}}}$$

$$\frac{\Delta[H_2O]}{\Delta t} = 2^* x \frac{-\Delta[CH_4]}{\Delta t} = \underline{\underline{0.32 \text{ mol } (l \cdot s)^{-1}}}$$

12.43 (a) $4NH_{3(g)} + 3O_{2(g)} \rightarrow 2N_{2(g)} + 6H_2O_{(g)}$ $\quad \Delta[N_2] = 0.68 \cdot \frac{mol}{l \cdot s}$

$$\frac{\Delta[H_2O]}{\Delta t} = 3^* x \frac{\Delta[N_2]}{\Delta t} = \underline{\underline{2.0 \text{ mol } (l \cdot s)^{-1}}}$$

(b) $\frac{\Delta[NH_3]}{\Delta t} = 2^* x \frac{-\Delta[N_2]}{\Delta t} = -1.4 \frac{mol}{l \cdot s}$

(c) $\frac{\Delta[O_2]}{\Delta t} = \frac{3}{2} x \frac{-\Delta[N_2]}{\Delta t} = \underline{\underline{-1.0 \text{ mol } (l \cdot s)^{-1}}}$

12.44 (a) rate = $2.35 \bullet 10^{-6} ltr^2 \bullet mol^{-2} \bullet s^{-1} (1 * mol \bullet ltr^{-1})^2 (1 * mol \bullet ltr^{-1})$

rate = $2.35 \bullet 10^{-6}$ mol/ltr•s (b) rate = $2.35 \bullet 10^{-6} (0.25)^2 (1.30)$ mol/ltr•s

= $1.9 \bullet 10^{-7}$ mol/ltr•s

12.45 $2A = 4B + C$

$$\text{slope(A)(25min)} = -\frac{0.38 mol \bullet l^{-1}}{38 \text{ min}}$$

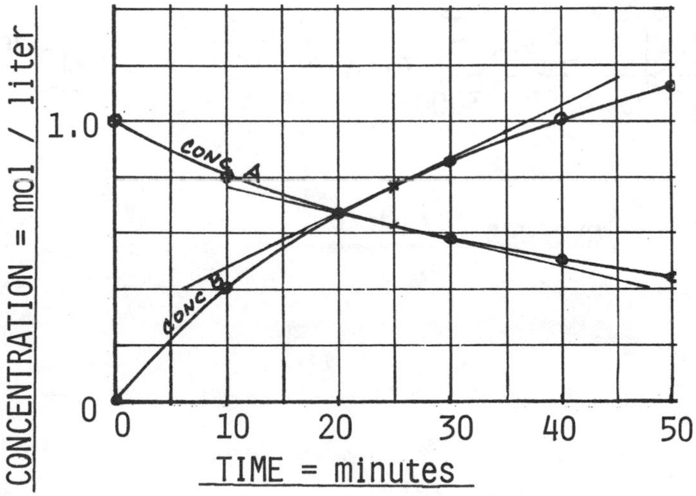

$$\text{slope(B)(25min)} = -\frac{0.77 mol \bullet l^{-1}}{39 \text{ min}}$$

$$\frac{\Delta[A]}{\Delta t} = \underline{-1.0 \times 10^{-2} \text{ mol}(l \bullet \text{min})^{-1}}$$

$$\frac{\Delta[B]}{\Delta t} = \underline{2.0 \times 10^{-2} \text{ mol}(l \bullet \text{min})^{-1}}$$

AT 40 MINUTES:

$$\frac{\Delta[A]}{\Delta t} = -\frac{0.28}{39} = \underline{-7.2 \times 10^{-3}}$$ $\Delta[B]/\Delta t = 0.54/40 = 1.4 \times 10^{-2}$

$\therefore \Delta[C]/\Delta t \simeq 5.0 \times 10^{-3}(25), 3.5 \times 10^{-3}(40)$

12.46 (a) rate = $1.63 \cdot 10^{-1}$ ltr/mol•s(0.25mol/ltr)(0.25mol/ltr) = $1.2 \cdot 10^{-2}$mol/ltr•s (b) rate = $2.0 \cdot 10^{-2}$mol/ltr•s (c) $4.1 \cdot 10^{-2}$mol/ltr•s

12.47

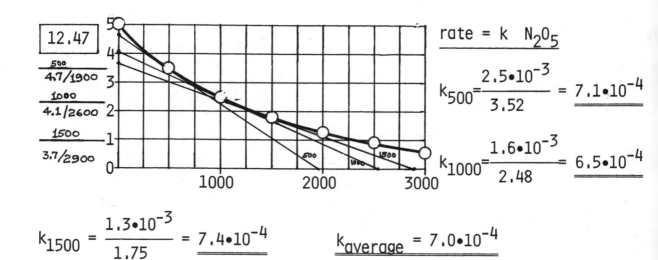

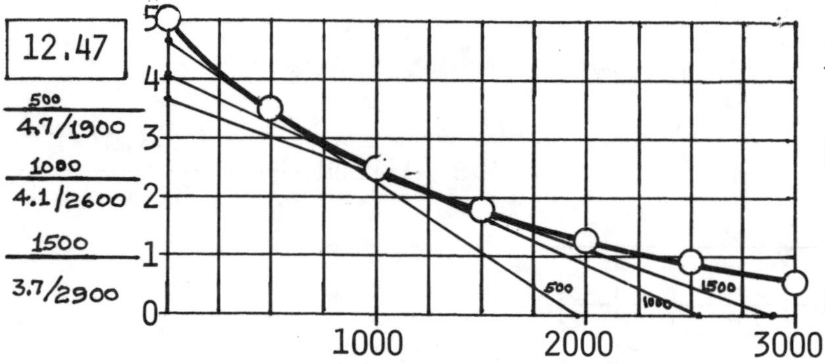

rate = k N_2O_5

$$k_{500} = \frac{2.5 \cdot 10^{-3}}{3.52} = 7.1 \cdot 10^{-4}$$

$$k_{1000} = \frac{1.6 \cdot 10^{-3}}{2.48} = 6.5 \cdot 10^{-4}$$

$$k_{1500} = \frac{1.3 \cdot 10^{-3}}{1.75} = 7.4 \cdot 10^{-4}$$

$k_{average} = 7.0 \cdot 10^{-4}$

12.48 (a) from data: (1) doubling $\{O_3\}$ doubles the rate

$\therefore$ rate $\propto$ $\{O_3\}$ (2) halving $\{NO_2\}$ halves rate $\therefore$ rate $\propto$ $\{NO_2\}$

$\therefore$ rate = k $\{NO_2\}$ $\{O_3\}$ k = rate / $\{NO_2\}$ $\{O_3\}$ =

$$\frac{0.022\ \text{mol/ltr•s}}{(5.0 \cdot 10^{-5})\text{mol/ltr}(1.0 \cdot 10^{-5})\text{mol/ltr}} = 4.4 \cdot 10^{7}\ \text{ltr/mol•s}$$

12.49 $\text{rate} = k\{NOCl\}^x$ Since doubling the concentration of NOCl increases the rate by four times and tripling the concentration of NOCl causes a ninefold increase in rate $\therefore$

$x = 2$. (a) $\text{rate} = k\{NOCl\}^2$ (b) $k = \dfrac{3.60 \cdot 10^{-9}\,\text{mol/ltr}\cdot\text{s}}{(0.30\,\text{mol/ltr})^2} = 4.0 \cdot 10^{-8}\ \dfrac{\text{ltr}}{\text{mol}\cdot\text{s}}$

(c) $(1.5)^2 = 2.25$

12.50 (a) $\text{rate} = k\{NO\}^2\{Cl_2\}$ (b) $k = \dfrac{2.53 \cdot 10^{-6}\,\text{mol/ltr}\cdot\text{s}}{(0.10\,\text{mol/ltr})^2(0.10\,\text{mol/ltr})}$

$= 2.5 \cdot 10^{-3}\,\text{ltr}^2/\text{mol}^2\cdot\text{s}$

12.51 For a first-order reaction the rate is proportional to $\{N_2O_5\}$.

(a) (eq. 12.6) $\ln \dfrac{[A]_o}{[A]_t} = kt \quad \therefore \{A\}_t = \{A\}_o\, e^{-kt} = 4.5 \cdot 10^{-3}\,\text{mol/ltr} \cdot$

$e^{-1.46 \cdot 10^{-1}\,\text{s}^{-1} \cdot 3600\,\text{s}} \simeq 0$ (N_2O_5 has almost all been consumed after 1hr)

(b) (eq. 12.8) $t_{1/2} = \dfrac{0.693}{k} = \dfrac{0.693}{1.46 \cdot 10^{-1}\,\text{s}^{-1}} = 4.75\text{s}$

(c) $\dfrac{1}{8}\{N_2O_5\} = 5.62 \cdot 10^{-4}\,\text{mol/ltr}$

12.52 $2HI \rightleftharpoons H_2 + I_2$ (second order)

$$t_{1/2} = \frac{1}{k\{HI\}_0} \quad \therefore \quad k = \frac{1}{t_{1/2}\{HI\}_0} = \frac{1}{2.11\,\text{min} \cdot 0.10\,\text{mol/ltr}} = \underline{4.74\ \text{ltr/mol} \cdot \text{min}}$$

(For greater precision an extra place is being carried in this intermediate answer. Only the final answer must be limited to 2 s.f.)

$$t_{1/2} = \frac{1}{4.74\ \text{ltr/mol} \cdot \text{min} \cdot 0.010\,\text{mol/ltr}} = \underline{\underline{21\ \text{min}}}$$

273
230
503

12.53 (see 12.52) $k = 4.74\ \text{ltr/mol} \cdot \text{min} \cdot \dfrac{1\ \text{min}}{60\ \text{s}} = \underline{\underline{0.079\ \text{ltr/mol} \cdot \text{s}}}$

12.54 $\underline{k(230^\circ C) = 0.163\ \text{ltr}^2/\text{mol}^2 \cdot \text{s}}$ $\underline{k(240^\circ C) = 0.348\ \text{ltr}^2/\text{mol}^2 \cdot \text{s}}$

$k(230^\circ C) = Ae^{-E_a/503K \cdot R}$ $k(240^\circ C) = Ae^{-E_a/513K \cdot R}$ $\therefore$ if the first equation is divided by the second $\therefore$ $\dfrac{0.163}{0.348} = \dfrac{Ae^{-E_a/503K \cdot R}}{Ae^{-E_a/513K \cdot R}} =$

$$e^{-E_a/503K \cdot R} \times e^{E_a/513K \cdot R} = 0.468$$

$$e^{(503E_a - 513E_a)/503 \cdot 513 \cdot R} = 0.468 \quad (R = 8.314\text{J/mol} \cdot \text{K})$$

$$\frac{-10E_a}{2.15 \cdot 10^6} = \ln 0.468 \qquad E_a = 0.759 \cdot 2.15 \cdot 10^6/10 = \underline{\underline{1.6 \cdot 10^5\ \text{J/mol}}}\ \underline{\underline{(160\ \frac{\text{kJ}}{\text{mol}})}}$$

(12.54 continued) $A = \frac{k}{e^{-E_a/RT}} = \frac{0.163\ \text{ltr}^2/\text{mol}^2\bullet\text{s}}{e^{-E_a/RT}} = 0.163\ e^{E_a/RT}$

$A = 1.36\bullet10^{16}\ \text{ltr}^2/(\text{mol}^2\bullet\text{s})$

12.55 (eq. 12.17) $\ln\frac{k_1}{k_2} = \frac{E_a}{R}\left(\frac{1}{T_2} - \frac{1}{T_1}\right)$ $\therefore$ $E_a = \frac{R\ \ln k_1/k_2}{(T_1-T_2)/T_1T_2}$

$$E_a = \frac{8.314\,\text{J}/(\text{mol}\bullet\text{K})\bullet\ln(1.32\bullet10^{-2}/1.64)}{-75\text{K}/(473\text{K}\bullet548\text{K})} = 1.39\bullet10^5\,\text{J/mol}\quad(139\,\text{kJ/mol})$$

(see 12.54) $A = \frac{1.64\ \text{ltr}/(\text{mol}\bullet\text{s})}{e^{-139{,}000\text{J/mol}(RT)}} = 2.91\bullet10^{13}\ \text{ltr}/(\text{mol}\bullet\text{s})$

12.56 (eq. 12.16) $\frac{k_1}{k_2} = e^{(E_a/R)((1/T_2) - (1/T_1))}$

$k_2 = 1.57\bullet10^{-3}\ \text{ltr}/(\text{mol}\bullet\text{s})\bullet e^{-(182{,}000\text{J}\bullet\text{mol}^{-1}/(8.314\text{J/mol}\bullet\text{K}))\bullet1.18\bullet10^{-4}\text{K}^{-1}}$

$= 1.19\bullet10^{-4}\ \text{ltr}/(\text{mol}\bullet\text{s})$

12.57 (from eq. 12.17) $\ln k_2 = \ln k_1 - \frac{E_a}{R}\left(\frac{1}{T_2} - \frac{1}{T_1}\right) = \ln(1.32\bullet10^{-2}) -$

$\frac{33.1\,\text{kcal/mol}}{1.987\bullet10^{-3}\,\text{kcal}/(\text{mol}\bullet\text{K})}\bullet\left(\frac{1}{573\text{K}} - \frac{1}{473\text{K}}\right) = 1.82$ $k_2 = e^{1.82}$

$k_2 = 6.16\ \text{ltr}/(\text{mol}\bullet\text{s})$

12.58 (eq. 12.17) $\ln(k_1/k_2) = (E_a/R)(1/T_2 - 1/T_1)$

$$E_a = \frac{\ln(k_1/k_2)}{(1/T_2 - 1/T_1)} \bullet R = \frac{(\ln 4^*) \bullet 8.314 \bullet 10^{-3}\,\text{kJ/(mol}\bullet\text{K)}}{6.194 \bullet 10^{-4}\,\text{K}^{-1}} = \underline{\underline{18.6\ \text{kJ/mol}}}$$

12.59

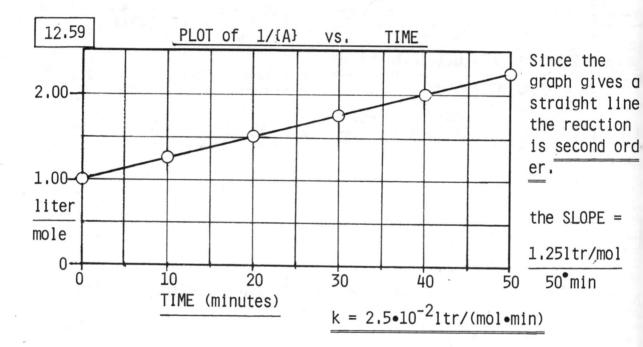

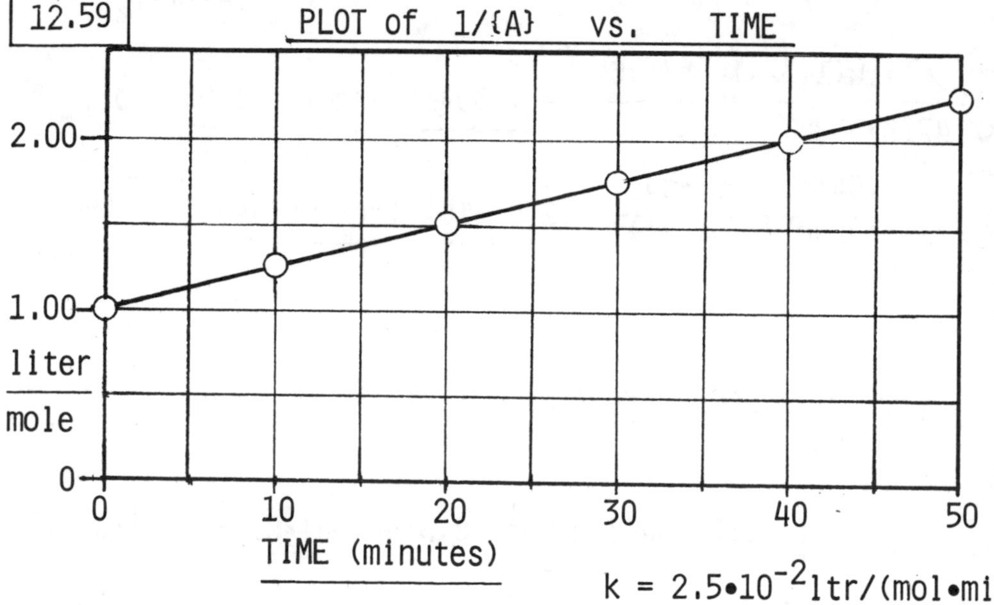

Since the graph gives a straight line the reaction is <u>second order</u>.

the SLOPE = $\dfrac{1.25\,\text{ltr/mol}}{50\bullet\text{min}}$

$\underline{\underline{k = 2.5 \bullet 10^{-2}\,\text{ltr/(mol}\bullet\text{min)}}}$

12.60 (from eq. 12.17)(see problem 12.58)

$$E_a = \frac{R \bullet \ln(k_1/k_2)}{(1/T_2 - 1/T_1)}$$

$$E_a = \frac{1.987 \bullet 10^{-3}\,\text{kcal/(mol}\bullet\text{K)} \bullet \ln(3.2 \bullet 10^{-2}/9.3 \bullet 10^{-2})}{(1.18 \bullet 10^{-3} - 1.22 \bullet 10^{-3})\,\text{K}^{-1}}$$

$\underline{\underline{E_a = 59\ \text{kcal/mol}}}$

12.61 (eq. 12.15) $\ln k = \ln A - E_a/RT$ $\qquad -\ln k = -\ln A + \frac{E_a}{R}\cdot\frac{1}{T}$

$1/T \cdot 10^3$	$-\ln k$
1.80	9.03
1.74	7.96
1.59	4.99
1.50	3.27
1.42	1.77

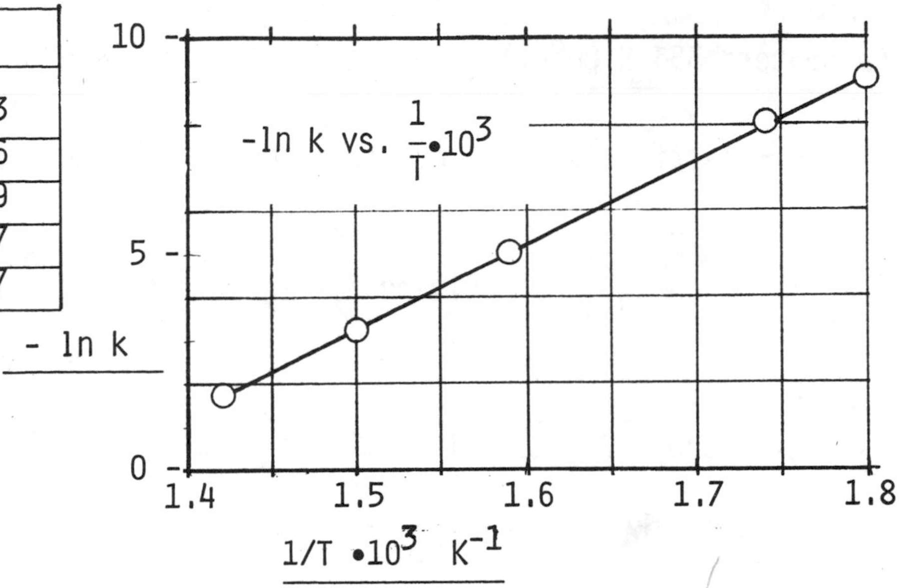

SLOPE = $1.9\cdot10^4$

E_a = SLOPE • R

$E_a = 1.6\cdot10^2$ kJ

12.62 NOTE: $k \propto 1/t$

$T^{-1}\cdot10^3$	t^{-1}	$\ln t^{-1}$
3.436	0.100	-2.303
3.413	0.111	-2.197
3.401	0.125	-2.079
3.390	0.143	-1.946
3.367	0.167	-1.792

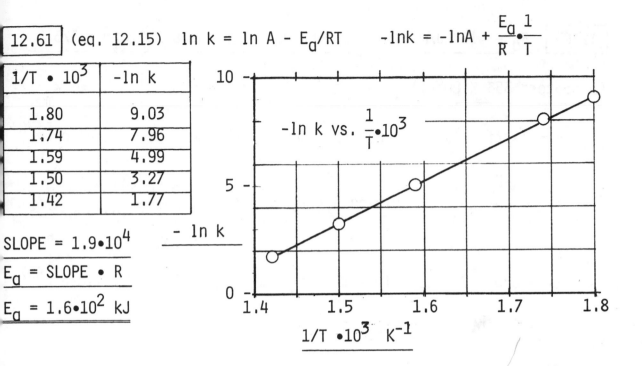

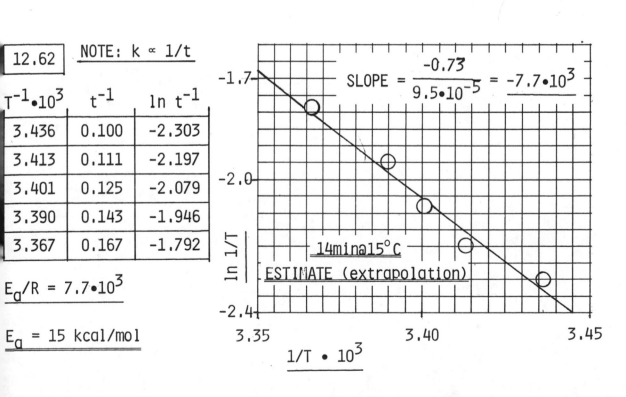

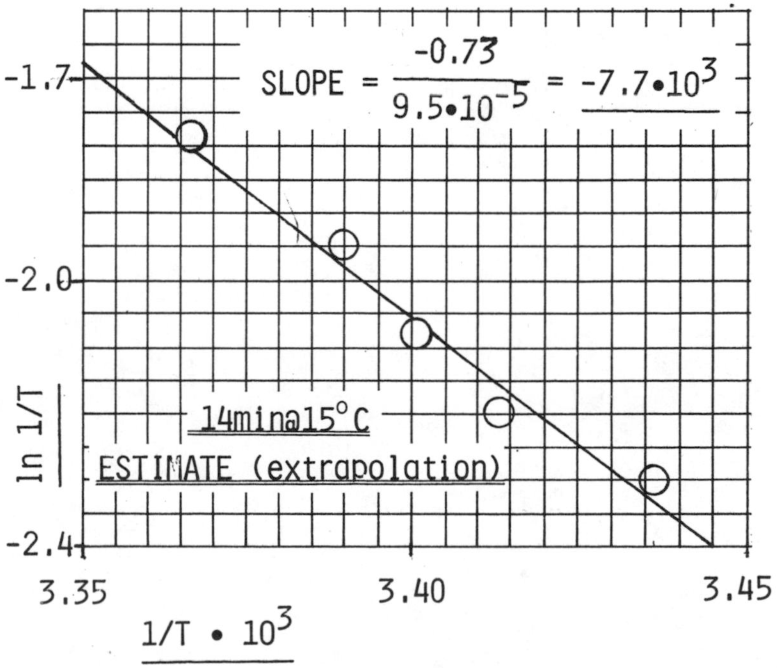

$E_a/R = 7.7\cdot10^3$

E_a = 15 kcal/mol

12.63 $E_a = 418$ kJ/mol   boiling point of H_2O @ 760torr = 373 K

(@ 355torr 353 K 80°C))   (eq. 12.14) $k = A\, e^{-E_a/RT}$ NOTE: $k \propto 1/t$

$\therefore$ $k = c \cdot 1/t$ ('c' is a constant of proportionality) $c/t = A\, e^{-E_a/RT}$

$$\frac{c/t_1}{c/t_2} = \frac{A\, e^{-E_a/RT_1}}{A\, e^{-E_a/RT_2}} \quad \text{(solving for } t_2\text{)} \quad t_2 = t_1 \cdot \frac{e^{-E_a/RT_1}}{e^{-E_a/RT_2}}$$

$$t_2 = t_1 \cdot e^{E_a/RT_2 - E_a/RT_1} = t_1 \cdot e^{E_a(T_1 - T_2)/RT_1T_2} = 3*\text{min} \cdot e^{7.637}$$

$t_2 = 6.2 \cdot 10^3$min (over 100 hours!)

NOTES FOR CHAPTER 13

The following notations will be used in the problem solutions for chapter 13: (1) "concentration of in moles per liter" will be indicated by either square brackets [ ] or { } (2) All concentrations listed in tables will be in MOLES PER LITER and in any case where concentration units are not specified MOLES PER LITER should be assumed. (3) The notation $\{\ \}_i$ represents "initial concentration of in moles per liter" while the notation $\{\ \}_{EQ}$ will be used to represent equilibrium concentrations.

Problems involving chemical equilibria can be confusing unless they are approached in a logical sequence. One useful technique utilizes a table. The first column indicates the species whose concentrations will be tabulated. Additional columns will show "initial" concentrations, "change" in concentrations which occur as the system attains equilibrium and the concentrations at equilibrium. In some cases the substitution of expressions which represent equilibrium concentrations by the use of algebraic unknowns into a mass action expression will lead to a quadratic or even a cubic equation. When this situation arises it is often possible to make simplifying assumptions which allow the equilibrium concentrations to be approximated in simpler terms which will not require the solution of a quadratic equation. The simplifying assumptions will be indicated in a column headed "assume:" and resulting approximate values of equilibrium concentrations will be listed in a column headed "approximation".

The technique used for the solution of some of the problems in this chapter assumes that if you know initial concentrations and the changes in concentrations which accompany the attainment of equilibrium, you can deduce the values of the concentrations at equilibrium. This logical process is implemented by the use of a concentration table.

PROBLEM SOLUTIONS FOR CHAPTER 13 of BRADY & HUMISTON 3rd Ed.

13.24 $K_c = \dfrac{[PCl_3][Cl_2]}{[PCl_5]} = \dfrac{(0.23)(0.055)}{(0.0023)} = \underline{5.5}$ (1)

5.6 (2), 5.4 (3), 5.5 (4) (Within experimental precision, all of these represent a $K_c = 5.5$

13.25 $K_c = \dfrac{[PCl_5]}{[PCl_3][Cl_2]} = \dfrac{1}{K_c(\text{prob. }13.24)} = \underline{0.18}$

13.26 $PCl_{5(g)} = PCl_{3(g)} + Cl_{2(g)}$ $K_c = 5.5\ mol \cdot l^{-1}$ (T = 298 K)

$\Delta n_{(g)} = +1$ $K_p = K_c(RT)^{\Delta n(g)} = 5.5\ \cancel{mol \cdot l^{-1}} \times 0.08205\ \cancel{l} \cdot atm \cdot \cancel{(mol \cdot K)^{-1}} \times$

$298\ \cancel{K} = \underline{1.3 \times 10^2}$

13.27 $K_c(773\ K) = 4.05$ $K_p = K_c(RT)^{\Delta n_{(g)}}$ $\Delta n_{(g)} = 0$ $\underline{K_p = 4.05}$

13.28 $K_c(1773\ K) = 5.67\ mol^2 \cdot l^{-2}$ (from stoichiometry) $\Delta n = +2$

$K_p = K_c(RT)^{\Delta n} = 5.67 \cdot \cancel{mol^2} \cdot \cancel{l^{-2}}(0.08205\ \cancel{l} \cdot atm \cdot (\cancel{mol} \cdot \cancel{K})^{-1} \cdot 1773\ \cancel{K})^2 =$

$\underline{1.20 \times 10^5 atm^2}$

13.29 $K_p(373\ K) = 6.5 \times 10^{-2} atm^{-1}$ (from stoichiometry)

$2\ NO_{2(g)} = N_2O_{4(g)}$ $K_c = K_p(RT)^{-\Delta n}$ $\Delta n = -1$ $K_c = 6.5 \cdot \times 10^{-2} \frac{1}{\cancel{atm}}$

$(0.08205\ l \cdot \cancel{atm} \cdot mol \cdot \cancel{K})^{-1} \cdot 373\ \cancel{K}) = \underline{2.0\ l \cdot mol^{-1}}$

13.30 $\Delta G' = -RT\ln K_p$ $K_p = e^{-\Delta G'/RT}$ $K_p = e^{3220 \cancel{cal \cdot mol^{-1}}/1.987 \cdot 700 \cancel{(cal/mol)}}$

$\underline{K_p = 10.13}$

13.31 $\Delta G' = -RT \ln K_P = -8.314 \cdot 10^{-3}$ kJ/mol$\cdot$K $\cdot 668$K$\cdot \ln(4.56 \cdot 10^{-2})$

$\Delta G'_{668K} = 17.1$ Kj/mol

13.32 $\Delta G' = -RT \ln K_P = -8.314 \cdot 10^{-3}$ kJ/(mol$\cdot$K)$\cdot 800$K$\cdot \ln(5.10)$

$\Delta G'_{800K} = -10.8$kJ/mol

13.33 $\Delta H^\circ = 0 + 0 -2(-92.5$ kJ/mol$) = 185$kJ/mol

$\Delta S^\circ = (130.6 + 223.0 -2(186.7))$J/(mol$\cdot$K) $= -19.8$J/(mol$\cdot$K)

$\Delta G'_{773} = \Delta H^\circ - T\Delta S^\circ = 185$kJ/mol -773K$(-1.98 \cdot 10^{-2}$kJ/(mol$\cdot$K)$) = 200$kJ/mol

$$\Delta G' = -RT \ln K_P \quad \therefore \quad \ln K_P = -\Delta G'/RT = \frac{-200 \text{ kJ/mol}}{8.314 \cdot 10^{-3} \text{ kJ} \cdot (\text{mol} \cdot \text{K})^{-1} \cdot 773\text{K}} =$$

$-31.2 \quad \therefore \quad K_P = 2.82 \cdot 10^{-14}$

13.34 $\Delta G^\circ = -RT \ln K_P = (-2(-22.8) - 0 + 2(-64.7) + 0)$kcal/mol $=$

-83.8kcal/mol $\quad \ln K_P = -\Delta G^\circ/RT \quad \therefore \quad K_P = e^{-\Delta G^\circ/RT} = e^{141.5} = 2.84 \cdot 10^{61}$

13.35 $\Delta G' = \Delta H^\circ - T\Delta S^\circ$, $\Delta H^\circ = (-20.2 - (12.4) - 0)\text{kcal/mol} = \underline{-32.6}$

$\Delta S^\circ = (54.9 - (52.5) - (31.21))\text{cal/mol} = \underline{-28.8\text{cal/mol}}$

$\Delta G' = (-32.6 - T(-28.8 \cdot 10^{-3}))\text{kcal/mol} = -RT \ln K_P = 0$ (when $K_P = 1$)

$28.8 \cdot 10^{-3} T = 32.6$ $\underline{T = 1130K\ (860°C)}$ (2 s.f.)

13.36 (eq. 13.4) $\Delta G = \Delta G^\circ + RT \ln(P_{CH_3OH}/(P_{H_2}^2 \cdot P_{CO}))$

$$\Delta G = -3.22\text{kcal} + 1.987 \cdot 10^{-3}(\text{kcal/(mol} \cdot K)) \cdot 700K \ln \frac{(3 \cdot 10^{-6})}{(1 \cdot 10^{-2})^2(2 \cdot 10^{-3})} =$$

$\underline{+0.547\text{kcal/mol}}$ (A positive value for ΔG means the system is not at equilibrium. The second term (+) must decrease untill it equals +3.22 for equilibrium to be attained (ΔG = 0). The second term will decrease when the numerator of the log term (P_{CH_3OH}) decreases and the denominator increases ($(P_{H_2})^2(P_{CO})$). This change represents decomposition of CH_3OH (spontaneous reaction to the left).

13.37 $P_{N_2O_5} = \underline{0.563 \text{ atm}}$ $\therefore P_{NO_2} = (0.844 - 0.563)\text{atm} = \underline{0.281 \text{ atm}}$

$$K_P = \frac{(P_{NO_2})^2}{(P_{N_2O_4})} = \frac{(0.281\text{atm})^2}{(0.563\text{atm})} = \underline{0.140 \text{ atm}} \text{ (a)}$$ (b) $K_c = K_P(RT)^{-\Delta n}$

$\Delta n = +1 \therefore \underline{K_c = 5.74 \cdot 10^{-3}\text{mol/ltr}}$ (c) $\Delta G^\circ = -RT \ln K_P = -8.314 \cdot 10^{-3}$ $(\text{kJ/(mol} \cdot K)) \cdot 298K \cdot \ln(0.140) = \underline{4.87\text{kJ/mol}}$

13.38 $\Delta G_f^o(SO_2) = \underline{-71.8 \text{ kcal/mol}}$ $\Delta G_f^o(NO_2) = \underline{12.4 \text{ kcal/mol}}$

$\Delta G_f^o(SO_3) = \underline{-88.5 \text{ kcal/mol}}$ $\Delta G_f^o(NO) = \underline{20.7 \text{ kcal/mol}}$

$\Delta G^o_{rctn} = \Delta G_f^o(NO) + \Delta G_f^o(SO_3) - \Delta G_f^o(SO_2) - \Delta G_f^o(NO_2)$

$\Delta G^o_{rctn} = 20.7\text{kcal/mol} - 88.5\text{kcal/mol} + 71.8\text{kcal/mol} - 12.4\text{kcal/mol} =$

$\underline{\underline{-8.4 \text{ kcal/mol}}}$ $\Delta G^o = -RT \ln K_P$ $K_P = e^{-\Delta G^o/RT} = e^{8.4 \cdot 10^3/298 \cdot 1.98}$

$= e^{14.186} = \underline{1.4 \cdot 10^6}$ $K_c = K_P(RT)^{-\Delta n}$ (Since $\Delta n = 0$, $\underline{\underline{K_c}} = K_P = \underline{\underline{1.4 \cdot 10^6}}$

13.39 (a) $\Delta G^o = \Delta G_f^o(NiSO_4) + 6 \cdot \Delta G_f^o(H_2O_{(g)}) - \Delta G_f^o(NiSO_4 \cdot 6H_2O)$

$\Delta G^o = (-189.4 + 6(-54.6) - (-531.0))\text{kcal/mol} = \underline{\underline{18.5\text{kcal/mol}}}$

(b) $K_P = e^{-\Delta G/RT} = e^{-18{,}500/1.987 \cdot 298} = e^{-31.24} = \underline{\underline{2.71 \cdot 10^{-14}}}$

(c) $K_P = (P_{H_2O})^6 = 2.71 \cdot 10^{-14}$ $\therefore$ $P_{H_2O} = \underline{5.48 \cdot 10^{-3}\text{atm}}$ $\underline{\underline{(4.16 \text{ torr})}}$

13.40 $K_c = \dfrac{\{PCl_3\}\{Cl_2\}}{\{PCl_5\}} = 33.3 \text{ mol/ltr}$ $PCl_3 = \dfrac{K_c \{PCl_5\}}{\{Cl_2\}}$

$\{PCl_3\} = 33.3\cancel{\text{mol/ltr}} \cdot (1.29 \cdot 10^{-3})(\text{mol/ltr}) / 1.87 \cdot 10^{-1} \cancel{\text{mol/ltr}} = \underline{\underline{0.230 \dfrac{\text{mol}}{\text{ltr}}}}$

13.41 $K_c = \frac{\{H_{2(g)}\}\{I_{2(g)}\}}{\{HI_{(g)}\}} = \frac{(1.0\cdot10^{-3})(2.5\cdot10^{-2})}{(2.2\cdot10^{-2})} = \underline{\underline{5.2\cdot10^{-2}}}$

13.42 $K_P = \frac{(P_{NOCl})^2}{P_{Cl_2}\cdot(P_{NO_2})^2} = \frac{(0.15)^2}{(0.18)(0.65)^2} = \underline{\underline{0.30\ atm^{-1}}}$

13.43 $\{O_2\}_i = 0.0560 mol/ltr \quad \{N_2O\}_i = 0.020 mol/ltr$

$\{NO_2\}_{EQ} = 0.020 mol/ltr \quad \therefore \quad \underline{\Delta\{N_2O\} = -2/4 \cdot 0.020 mol/ltr} \quad \therefore \quad \{N_2O\}_{EQ} =$

$\{N_2O\}_i + \Delta\{N_2O\} = (0.020 - 0.010) mol/ltr = \underline{\underline{0.010 mol/ltr}}$ (a)

$\Delta\{O_2\} = -3/4 \cdot 0.020 mol/ltr = \underline{-0.015 mol/ltr} \quad \therefore \quad \{O_2\}_{EQ} = \{O_2\}_i + \Delta\{O_2\} =$

$0.0560 mol/ltr - 0.015 mol/ltr = \underline{\underline{0.041\ mol/ltr}}$ (a)

(b) $\underline{\underline{K_c}} = \frac{\{NO_2\}^4}{\{N_2O\}^2\{O_2\}^3} = \frac{(0.020)^4}{(0.010)^2(0.041)^3} = \underline{\underline{23}}$

13.44 The construction of a table showing the the species involved in the problem, with the initial concentrations, the changes in concentrations and the concentrations at EQUILIBRIUM, is often a very valuable aid to the problem solving process. A table of this type is utilized for the solution of this problem.

(13.44 continued)

$$K_c = \frac{\{NO\}\ \{SO_3\}}{\{SO_2\}\ \{NO_2\}} = 85.0$$

conc.	initial	change	EQUILIBRIUM
SO_2	0.0500	-X	0.0500 - X
NO_2	0.0500	-X	0.0500 - X
NO	0	+X	X
SO_3	0	+X	X

$$\frac{(X)(X)}{(0.0500 - X)^2} = 85.0$$

(We can avoid solving a quadratic equation by taking the square root of both sides of the above equation.)

$$\frac{X}{0.0500 - X} = 9.22$$

$X = 0.461 - 9.22\ X$ $\qquad$ $\underline{X = 0.0451}$

$\therefore$ $\{SO_2\}_{EQ} = 0.0049 = \{NO_2\}_{EQ}$ $\qquad$ $\{NO\}_{EQ} = 0.0451 = \{SO_3\}_{EQ}$

13.45 $\quad K_c = \frac{\{CO\}\ \{H_2O\}}{\{H_2\}\ \{CO_2\}} = 0.771$ let X = the change to attain equilib

$\{H_2\} = \{CO_2\} = (0.200 - X)$ $\qquad$ $\{CO\} = \{H_2O\} = X$ $\qquad$ $\frac{(X)^2}{(0.200-X)^2} = 0.771$

(see 13.44) $\quad \frac{X}{(0.200-X)} = 0.878$ $\qquad$ $X = 0.878(0.200-X)$ $\qquad$ $\underline{X = 0.0935}$

$\{CO\} = \{H_2O\} = 0.0935\frac{mol}{ltr}$ $\quad$ $\{H_2\} = \{CO_2\} = (0.200 - 0.0935) = 0.106 mol/ltr$

13.46 $\quad K_c = \frac{\{NO\}\ \{SO_3\}}{\{NO_2\}\ \{SO_2\}} = 85.0 = \frac{(0.0100 + X)(0.0150 + X)}{(0.0200 - X)(0.0100 - X)}$ (SEE TABLE NEXT PAGE)

(13.46 continued)

$$= \frac{1.50 \bullet 10^{-4} + 0.0250\,X + X^2}{2.00 \bullet 10^{-4} - 0.0300\,X + X^2}$$

$$84.0X^2 - 2.575X + 0.01685 = 0$$

conc.	initial	change	EQUILIBRIUM
SO_2	0.0100	-X	0.0100 - X
NO_2	0.0200	-X	0.0200 - X
NO	0.0100	+X	0.0100 + X
SO_3	0.0150	+X	0.0150 + X

SOLUTION OF QUADRATIC EQUATION:

0.0211868 & 0.0094679 ← (this is the physically significant solution)
(not possible)

IF X = 0.0094679 ∴ $\{SO_2\}_{EQ}$ = 0.0100 - 0.0094679 = 0.0005 $\{NO_2\}_{EQ}$ = 0.0200 - 0.0094679 = 0.0105 $\{NO\}_{EQ}$ = 0.0100 + 0.0094679 = 0.0195 $\{SO_3\}_{EQ}$ = 0.0150 + 0.0094679 = 0.0245

Although some electronic calculators allow the direct solution of quadratic equations the algebraic manipulations involved are a frequent source of error.

There is an alternative method for the solution of this problem which recognizes that an equilibrium constant of 85 indicates that almost all of the initial concentration of SO_2 would be consumed in the change step. (X ≈ 0.010) Based on this assumption, the equilibrium concentrations would be: $\{SO_2\}$ ≈ very small, $\{NO_2\}$ ≈ 0.01, $\{NO\}$ ≈ 0.02 $\{SO_3\}$ ≈ 0.025 If these approximations are substituted into the equilibrium constant expression, the concentration of SO_2 is the only remaining unknown quantity.

$$\frac{(0.020)(0.025)}{(0.010)\ \{SO_2\}} = 85.0$$

$\{SO_2\}$ ≈ 0.0006

(Note that all of these values are close to those obtained by the use of the quadratic equation.)

Since X = $-\Delta\ SO_2$ (which equals) $SO_{2\,i}$ - $SO_{2\,EQ}$ = 0.0100 - 0.0006 = 0.0094 This value for X is more precise than the assumption that X = 0.01 so substitution of 0.0094 for X in the table above will yield the following values for EQUILIBRIUM concentrations: $\{NO_2\}_{EQ}$ = 0.0106M $\{NO\}_{EQ}$ = 0.0194M $\{SO_3\}_{EQ}$ = 0.0244M (These values are very close to the quadratic solutions and can be used to calculate $\{SO_2\}$ = 0.0005 .

(13.46 continued) These values for the EQUILIBRIUM concentrations are as satisfactory as those obtained by the solution of the quadratic equation.

It is often possible (as in this example) to make a simplfying assumption which allows the calculation of fairly precise answers by simple arithmetic. (More precise answers may be obtained by successive approximations. In this example only a second approximation was required to equal the precision of the quadratic solution.)

13.47 $K_c = \dfrac{\{CO\}^2\,\{O_2\}}{\{CO_2\}^2} = 6.4 \cdot 10^{-7}$ let $X = -\Delta\{CO_2\}$

$$\frac{(X)^2(X/2)}{(1 \cdot 10^{-3} - X)^2} = 6.4 \cdot 10^{-7}$$

conc.	initial	change	EQUILIBRIUM
CO_2	$1.0 \cdot 10^{-3}$	$-X$	$1.0 \cdot 10^{-3} - X$
CO	0	$+X$	X
O_2	0	$+ X/2$	$X/2$

In the solution to 13.46 the idea of simplifying assumptions was offered as an alternative to the solution of a quadratic equation. In this problem the exact solution involves a cubic equation and the simplifying assumption method is a much easier alternative!

The very small value of the K_c indicates that the equilibrium lies far to the left, i.e. very little CO_2 is used up in the change step. ($X \ll 1.0 \cdot 10^{-3}$) If X is very small compared to $1.0 \cdot 10^{-3}$, the EQUILIBRIUM concentration of $CO_2 \simeq 1.0 \cdot 10^{-3}$. (As demonstrated below, this approximation yields an equation which can be solved by taking the cube root of both sides, a far less complicated procedure than the solution of the previous cubic equation.)

$$\frac{(X)^2(X/2)}{(1 \cdot 10^{-3})^2} = 6.4 \cdot 10^{-7} \qquad X^3/2 = 6.4 \cdot 10^{-13} \qquad X^3 = 12.8 \cdot 10^{-13}$$

$\underline{X = 1.1 \cdot 10^{-4}}$ (Although this value is 11% of $1 \cdot 10^{-3}$ which is not 'very small' the EQUILIBRIUM concentrations based on this assumption are calculated below:

$\{CO_2\}_{EQ} = 1.0 \cdot 10^{-3} - 1.1 \cdot 10^{-4} = \underline{0.0009M}$, $\{CO\}_{EQ} = \underline{0.0001M}$, $\{O_2\}_{EQ} = \underline{0.00006M}$

How much uncertainty is associated with these values due to the approximation employed in the calculation? In order to answer this question we will make a second approximation to increase

(13.47 continued) the precision of our answers and also compare the results with answers obtained by solving the cubic equation. <u>second approximation:</u>

$$\frac{(X)^2(X/2)}{(9\cdot10^{-4})^2} = 6.4\cdot10^{-7}$$

$X^3 = 1.04\cdot10^{-12}$ $\underline{X = 1.0\cdot10^{-4}}$ (This value is close to that obtained by the first approximation. The concentrations calculated using this value of X are shown below.)

$\{CO_2\}_{EQ} = 1.0\cdot10^{-3} - 1.0\cdot10^{-4} = \underline{0.0009}$ (EXACTLY THE SAME ANSWER TO THE 1 s.f. ALLOWED BY THE DATA!)

Since the precision of the data allows only 1 s.f. in the answer, no improvement resulted from the second approximation. Can more precise answers be calculated by an exact solution to the cubic equation?

$X^3/2 = 6.4\cdot10^{-7}X^2 - 1.28\cdot10^{-9}X + 6.4\cdot10^{-13}$ (the only 'real'root)

$0.5X^3 - 6.4\cdot10^{-7}X^2 + 1.28\cdot10^{-9}X - 6.4\cdot10^{-13} = 0$ $\underline{X = 1.01\cdot10^{-4}}$

The solution of the cubic equation yields exactly the same value for X as the second approximation(to 3 s.f.). In this problem the 2 s.f. of data allow only 1 s.f. in the answer, and that fact rather than the method used determines the precision of the final answer. Since the data used for problems involving EQUILIBRIUM is often the limiting factor, approximation techniques often allow calculation of an answer which is just as satasfactory as an exact solution. (b) $\underline{\text{fraction} = 0.10}$

13.48 let $X = -\Delta\{COCl_2\}$

conc.	initial	change	EQUILIBRIUM
$COCl_2$	0.020	- X	0.020-X
CO	0	+ X	X
Cl_2	0	+ X	X

$$\frac{\{COCl_2\}}{\{CO\}\ \{Cl_2\}} = 4.6\cdot10^9$$

$$\frac{(0.020-X)}{(X)\ (X)} = 4.6\cdot10^9$$

(see 13.47) <u>ASSUME: $X \ll 0.020$</u> $(0.020-X) \simeq 0.020$

$0.020 = 4.6\cdot10^9X^2$ $\underline{X = 2.09\cdot10^{-6} = \{CO\} = \{Cl_2\}}$ $\underline{\{COCl_2\} = 0.020}$

13.49 $K_P = P_{CO_2} \bullet P_{H_2O} = 0.25\ atm^2$ $P_{CO_2} = P_{H_2O} = \sqrt{0.25\ atm^2} = 0.50 atm$

The carbon dioxide is liberated in small bubbles causing the dough to rise before baking.

13.50 $2HI \rightleftharpoons H_2 + I_2$ (assume all species are gases)

$$K_c = \frac{\{H_2\}\{I_2\}}{\{HI\}^2} = \frac{(0.0100)(0.0100)}{(0.0740)^2} = 1.83 \bullet 10^{-2}$$

The addition of 0.50 mol of HI (0.050mol/ltr) will upset the EQUILIBRIUM.

$$\text{mass action expression} = \frac{(0.0100)^2}{(0.1240)^2} = 6.50 \bullet 10^{-3} < K_c \quad (1.83 \bullet 10^{-2})$$

To return to an equilibrium condition some of the added HI must decompose to form additional H_2 and I_2. let $-X = \Delta\{HI\}$ $\therefore \{HI\}_{EQ} = (0.1240-X)$

$\{H_2\}_{EQ} = (0.0100 + X/2) = \{I_2\}_{EQ}$ $\quad \frac{(0.0100+X/2)^2}{(0.1240-X)^2} = 1.83 \bullet 10^{-2}$ ($\sqrt{\ }$ both)

$(0.0100 + X/2) = (0.1240 - X) \bullet 0.135$ $\quad 0.5*X + 0.135X = 0.1240 \bullet 0.135 - 0.0100$ $\quad 0.635X = 0.00674$ $\quad X = 0.0106$ $\therefore \{HI\}_{EQ} = 0.1240 - 0.0106 = 0.113\ M$ $\quad \{H_2\}_{EQ} = 0.0100 + 0.0106/2 = 0.0153 = \{I_2\}_{EQ}$

13.51 let $X = -\Delta\{CO\}$ (This problem solution will utilize a table which includes a simplifying assumption concerning the magnitude of X, and the EQUILIBRIUM concentrations as approximations based on that assumption.)

(13.51 continued) $CO_{(g)} + Cl_{2(g)} \rightleftharpoons COCl_{2(g)}$

$\frac{\{COCl_2\}}{\{CO\}\{Cl_2\}} = K_c = 4.6 \cdot 10^9$

conc.	initial	change	EQUILIBRIUM	assume:	approximation
CO	0.15	-X	0.15 - X	X≃0.15	(very small)
Cl_2	0.30	-X	0.30 - X		0.15
$COCl_2$	0	+X	X		0.15

using the approximations: $\frac{(0.15)}{\{CO\}(0.15)} = 4.6 \cdot 10^9 \therefore \{CO\} = 2.2 \cdot 10^{-10} M$

$\{Cl_2\} = \{COCl_2\} = 0.15M$ (Since the difference between 0.15 and X is equal to {CO} ($2.2 \cdot 10^{-10}$) our assumption that X≃0.15 was valid.)

13.52 $N_{2(g)} + O_{2(g)} \rightleftharpoons 2NO_{(g)}$ $K_P(1000°C) = 4.8 \cdot 10^{-7}$

PRESS.	INITIAL	CHANGE	EQUILIBRIUM
N_2	33.6	- X	33.6 - X
O_2	4.0	- X	4.0 - X
NO	0	+2X	2X

LET $-X = \Delta P_{(N_2)}$ $(= \Delta P_{(O_2)})$

$$K_P = \frac{(P_{NO})^2}{P_{(N_2)} \cdot P_{(O_2)}} = 4.8 \cdot 10^{-7}$$

$$\frac{(2X)^2}{(33.6-X)(4.0-X)} = 4.8 \cdot 10^{-7}$$

(Since K_P is small, ASSUME the change to attain EQUILIBRIUM (X) is small compared to the starting pressure of oxygen (4.0). (X<<4.0) The mass action expression can then be reduced to:

$$\frac{(2X)^2}{(33.6)(4.0)} = 4.8 \cdot 10^{-7} \qquad 4X^2 = 6.45 \cdot 10^{-5}$$

X = 0.0040 (ASSUMPTION VALID) $P_{(NO)} = 2X = 0.0080$ atm

13.53 $\frac{P_i(NO)}{P_i(N_2)} = \frac{P_f(NO)}{P_f(N_2)}$ $\therefore$ $P_f(NO) = 8.0 \cdot 10^{-3} atm(i) \frac{0.80\ atm(f)}{33.6\ atm(i)} =$

$1.9 \cdot 10^{-4} atm$

13.54 $2NO_2 \rightleftharpoons N_2O_4$ $K_c = \frac{\{N_2O_4\}}{\{NO_2\}^2} = 7.5$ let $X = \Delta\{N_2O_4\}$

conc.	initial	change	EQUILIBRIUM
NO_2	1.0	-2X	1.0-2X
N_2O_4	0	+ X	X

$$7.5 = \frac{X}{(1.0-2X)^2}$$

$$(4X^2 - 4X + 1.0) \cdot 7.5 = X$$

$$30X^2 - 31X + 7.5 = 0$$

(this value of X makes $\{NO_2\} = (-)$

roots = 0.6468 and 0.3865 ←←←(the physically significant root)

$\{NO_2\} = 1.0 - 2 \cdot (0.3865) = 0.2$ M $\{N_2O_4\} = 0.39$ M

+++++IN A DOUBLE SIZED CONTAINER+++++

conc.	initial	change	EQUILIBRIUM
NO_2	0.50	-2X	0.50-2X
N_2O_4	0	+ X	X

$$7.5 = \frac{X}{(0.50-2X)^2}$$

$$(4X^2 - 2X + 0.25) \cdot 7.5 = X$$

$$30X^2 - 16X + 1.875 = 0$$

roots = 0.3595 and 0.1739 ←←←(the physically significant root)

$\{NO_2\} = 0.50 - 2X = 0.15$ $\{N_2O_4\} = 0.17$

The equilibrium in the 2 liter container involved a concentration of N_2O_4 which was almost twice the concentration of NO_2. In the 4 liter container the concentration of NO_2 was proportionally larger. This is in line with the Le Chatelier prediction. Larger volume favors more ga

NOTES FOR CHAPTER 15

(Please reread the notes which precede the problem solutions for chapter 13 since they also apply to chapter 15.) Since this chapter deals with acid-base equilibria in aqueous solution some additional information is required to understand these problem solutions. In some of these problems additional columns will be used in the concentration table as follows: When more than a single ionization stage is involved each will be treated in a separate "change" column. The symbol >> will be used to signify "very much larger than". EXAMPLE: If X>>Y then X + Y ≈ X and X - Y is also ≈ X.

When there is more than one source of an ion in a system it is often the case that one of the sources provides so much larger concentration of the ion than any of the others that the equilibrium concentration of that ion can be closely approximated as that of the large source alone. This is frequently the case when a weak acid is present in water solution and hydronium ion is supplied by the ionization of water and also by the ionization of the weak acid. If the weak acid is present in appreciable concentration, the hydronium ion from the ionization of that weak acid is usually much larger than the hydronium ion from the ionization of H_2O (10^{-7}) and the 10^{-7} can be ignored. In the concentration tables used in the solutions which follow, however, the 10^{-7}mol/liter of H^+ (or OH^-) from water is always included as an initial concentration and is only discarded on the basis of a valid assumption that it is small enough to neglect in comparison to some other source. If you fall into the habit of not even considering the H^+ and OH^- from the ionization of water, you occasionally calculate an incorrect answer! (As an example of such a problem see *15.22.) The answer to 15.22 is NOT that a 10^{-8}M HCl solution is slightly basic with a pH of 8!

PROBLEM SOLUTIONS FOR CHAPTER 15 of BRADY & HUMISTON 3rd Ed.

15.19 (a) 10^{-3}M HCl = $[H^+] = 10^{-3}$ $[OH^-] = 10^{-11}$ pH = 3.0

(b) $[H^+] = 0.125M$ $[OH^-] = 8.00 \times 10^{-14}$ pH = 0.903

(c) $[OH^-] = 0.0031M$ $[H^+] = 3.2 \times 10^{-12}$ pH = 11.49

(d) $[OH^-] = 0.024M$ $[H^+] = 4.2 \times 10^{-13}$ pH = 12.38 (COMPLETE DISSOCIATION ASSUMED)

(e) $[H^+] = 2.1 \times 10^{-4}M$ $[OH^-] = 4.8 \times 10^{-11}$ pH = 3.68

(f) $[H^+] = 1.3 \times 10^{-5}M$ $[OH^-] = 7.7 \times 10^{-10}$ ph = 4.89

(g) $[OH^-] = 8.4 \times 10^{-3}M$ $[H^+] = 1.2 \times 10^{-12}$ pH = 11.92

(h) $[OH^-] = 4.8 \times 10^{-2}M$ $[H^+] = 2.1 \times 10^{-13}$ pH = 12.68

15.20 (a) $[H^+] = 0.050M$ $[OH^-] = 2.0 \times 10^{-13}M$

(b) $1.9 \times 10^{-6}M$, $5.4 \times 10^{-9}M$ (c) $1.0 \times 10^{-4}M$, $1.0 \times 10^{-10}M$

(15.20 continued) (d) 1.6×10^{-8}M, 6.3×10^{-7}M

(e) 1.1×10^{-11}M, 8.7×10^{-4}M (f) 2.5×10^{-13}M, 4.1×10^{-2}M

15.21 $pOH = -\log [OH^-]$ (a) 12.70 (b) 8.27 (c) 10.00

(d) 6.20 (e) 3.06 (f) 1.39 ALSO: $pOH = 14.00 - pH$

15.22

CONC.	H_2O	ADD	EQUIL.
H^+	$1.0 \cdot 10^{-7}$	$1.0 \cdot 10^{-8}$	$1.0 \cdot 10^{-7} + 1.8 \cdot 10^{-8}$

$\therefore$ $H^+ = 1.1 \cdot 10^{-7}$

pH = 6.96 (very slightly acidic - see discussion of Ch 15 problm soltns)

15.23 $pK_a = -\log K_a = 8.42$

15.24 $K_b = 10^{-pK_b} = 1.4 \cdot 10^{-4}$

15.25 (a) $HNO_2 \rightleftharpoons H^+ + NO_2^-$ let $X = \Delta\{H^+\}$

conc.	initial	change	EQUILIBRIUM	assume	∴
H^+	$1\cdot10^{-7}$	+ X	$1\cdot10^{-7}$ + X	$X >> 10^{-7}$	X
NO_2^-	0	+ X	X	$X << 0.3$	X
HNO_2	0.30	- X	0.30 - X		0.3

$$\frac{\{H^+\}\{NO_2^-\}}{\{HNO_2\}} = 4.5\cdot10^{-4}$$

$$\frac{X^2}{0.30} = 4.5\cdot10^{-4}$$

X = 0.012M = $\{H^+\}$ (both assumptions were satisfactory)

(b) $2.6\cdot10^{-2}$M (c) $3.5\cdot10^{-6}$M (d) $1.2\cdot10^{-3}$M (e) $7.1\cdot10^{-4}$M

15.26 (a) $NH_3 + H_2O \rightleftharpoons NH_4^+ + OH^-$ let $X = \Delta\{OH^-\}$

conc.	initial	change	EQUILIBRIUM	assume:	∴
OH^-	$1\cdot10^{-7}$	+ X	$1\cdot10^{-7}$ + X	$X >> 10^{-7}$	X
NH_3	0.15	- X	0.15 - X	$X << 0.15$	0.15
NH_4^+	0	+ X	X		X

$$\frac{\{NH_4^+\}\{OH^-\}}{\{NH_3\}} = K_b$$

$K_b = 1.8\cdot10^{-5}$

$$\frac{(X)(X)}{0.15} = 1.8\cdot10^{-5}$$

$X^2 = 2.7\cdot10^{-6}$ $X = 1.6\cdot10^{-3}$ NOTE: 0.15>>0.0016

$0.0016 >> 10^{-7}$ ∴ (both of the assumptions were valid) $\{OH^-\}$ = X = 0.0016

(b) $\{OH^-\} = 5.8\cdot10^{-4}$ (c) $\{OH^-\} = 1.7\cdot10^{-2}$ (d) $6.2\cdot10^{-5}$ (e) $4.2\cdot10^{-6}$

15.27 $\{OH^-\} = 10^{-14}/\{H^+\}$ (a) $\{OH^-\} = 8.3\cdot10^{-13}$M (b) $3.8\cdot10^{-13}$M

(15.27 continued) (c) $\{OH^-\} = 2.9 \cdot 10^{-9} M$ (d) $\{OH^-\} = 8.3 \cdot 10^{-12} M$
(e) $\{OH^-\} = 1.4 \cdot 10^{-11} M$

15.28 pH = 14.00 - pOH (a) pOH = 2.80, pH = 11.20 (b) pOH = 3.24, pH = 10.76 (c) pOH = 1.77, pH = 12.23 (d) pOH = 4.21, pH = 9.79 (e) pOH = 5.38, pH = 8.62

15.29 pH = 1.35 $\{H^+\} = 10^{-pH}$ $\{H^+\}_{EQ} = 0.045M$ weak acid = HA

conc.	initial	change	EQUILIBRIUM	since;	∴
HA	0.25	- X	0.25 - X	X=0.045	0.21
H^+	10^{-7}	+ X	10^{-7} + X		0.045
A^-	0	+ X	X		0.045

$$\frac{\{H^+\}\{A^-\}}{\{HA\}} = K_a \qquad \frac{(0.045)(0.045)}{(0.21)} =$$

$K_a = 9.6 \cdot 10^{-3}$ (NOTE: $X = \Delta\{H^+\}$)

15.30 pH = 5.37 $\{H^+\}_{EQ} = 4.3 \cdot 10^{-6} M$ (See 15.29 for details; this problem is identical except for the numerical values.)

$\{A^-\}_{EQ} = \{H^+\}_{EQ} = 4.3 \cdot 10^{-6} M$ $\{HA\}_{EQ} = \{HA\}_i + \Delta\{HA\} = 0.10 - 4.3 \cdot 10^{-6}$

$= 0.10M$ $(4.3 \cdot 10^{-6})^2 / 0.10 = K_a = 1.8 \cdot 10^{-10}$

15.31 pH = 8.75 ∴ $\{H^+\} = 1.8 \cdot 10^{-9}M$, $\{OH^-\} = 5.6 \cdot 10^{-6}M$

conc.	initial	change	EQUILIBRIUM	since:	∴
MOH	0.10	- X	0.10 - X	$X = 5.6 \cdot 10^{-6}$	0.10
M^+	0	+ X	X		$5.6 \cdot 10^{-6}$
OH^-	10^{-7}	+ X	10^{-7} - X		$5.6 \cdot 10^{-6}$

weak base = MOH

$$\frac{\{OH^-\}\{M^+\}}{\{MOH\}} = K_b$$

$$K_b = \frac{(5.6 \cdot 10^{-6})^2}{0.10} = 3.2 \cdot 10^{-10}$$

(NOTE: $X = \Delta\{OH^-\}$)

15.32 (a) $HCO_2H \rightleftharpoons H^+ + CO_2H^-$ $\quad K_a = \frac{\{H^+\}\{CO_2H^-\}}{\{HCO_2H\}} = 1.8 \cdot 10^{-4}$

conc.	initial	change	EQUILIBRIUM	assume:	∴
HCO_2H	1.0	- X	1.0 - X	$X \ll 1$	1.0
H^+	10^{-7}	+ X	10^{-7} + X	$X \gg 10^{-7}$	X
CO_2H^-	0	+ X	X		X

let $X = \Delta\{H^+\}$

$$\frac{(X)(X)}{1.0} = K_a$$

$$X^2 = 1.8 \cdot 10^{-4}$$

$X = 0.013M = H^+$ $\quad$ % ionization = X/1.0 x 100 = 1.3%

(b) 3.7%ionized (c) 0.014% ionized (d) 0.63% ionized (e) 0.025%

(f) assume 100% ionized

15.33 $HC_2H_3O_2 \rightleftharpoons H^+ + C_2H_3O_2^-$ $\quad K_a = 1.8 \cdot 10^{-5} = \frac{\{H^+\}\{C_2H_3O_2^-\}}{\{HC_2H_3O_2^-\}}$

In the above equilibrium the rate of the reverse reaction depends upon the frequency of collisions between H^+ and $C_2H_3O_2^-$ ions. In a more

(15.33 continued)dilute solution the frequency of collisions between these ions decreases and a greater extent of ionization is necessary to establish equilibrium. This point is illustrated by the results obtained in this problem solution.

conc.	initial	change	EQUILIBRIUM	asuume:	∴
$HC_2H_3O_2$	1.00	- X	1.00 - X	X<<1	1.00
H^+	10^{-7}	+ X	10^{-7} + X	X>>10^{-7}	X
$C_2H_3O_2^-$	0	+ X	X		X
$HC_2H_3O_2$	0.10	- X	0.10 - X	X<<0.1	0.10
H^+	10^{-7}	+ X	10^{-7} + X	X>>10^{-7}	X
$C_2H_3O_2^-$	0	+ X	X		X
$HC_2H_3O_2$	0.010	- X	0.010-X	X<<0.01	0.010
H^+	10^{-7}	+ X	10^{-7} + X	X>>10^{-7}	
$C_2H_3O_2^-$	0	+ X	X		X

$X = \Delta\{H^+\}$

(a) $X = 4.2 \cdot 10^{-3}$
(assumption valid)
∴ 0.42% ioniz.

(b) $X = 1.3 \cdot 10^{-3}$
(assumption valid)
∴ 1.3% ioniz.

(c) $X = 4.2 \cdot 10^{-4}$
(assumption valid)
∴ 4.2% ioniz.

NOTE THAT THE EXTENT OF IONIZATION INCREASED WITH DILUTION.

15.34 (a) $HC_2H_3O_2 \rightleftharpoons H^+ + C_2H_3O_2^-$ $\qquad K_a = \dfrac{\{H^+\}\{C_2H_3O_2^-\}}{\{HC_2H_3O_2\}}$

conc.	initial	change	EQUILIBRIUM	assume:	∴
$HC_2H_3O_2$	0.25	- X	0.25 - X	X<<0.25	0.25
H^+	10^{-7}	+ X	10^{-7} +XX	X>>10^{-7}	X
$C_2H_3O_2^-$	0.15	+ X	0.15 + X	X<<0.15	0.15

$$\frac{X(0.15)}{0.25} = 1.8 \cdot 10^{-5}$$

$X = 1.1 \cdot 10^{-5}$

$\{H^+\} = 1.1 \cdot 10^{-5}$

(15.34 continued) (b) $HCHO_2 \rightleftharpoons H^+ + CHO_2^-$ $K_a = \frac{\{H^+\}\{CHO_2^-\}}{\{HCHO_2\}}$

since: $\{CHO_2^-\} = \{HCHO_2\} = 0.50M$ $\therefore$ $\{H^+\} = K_a = 1.8 \cdot 10^{-4}$

(c) $\frac{\{H^+\}\{NO_2^-\}}{\{HNO_2\}} = K_a = 4.5 \cdot 10^{-4} = \frac{\{H^+\}(0.40)}{(0.30)}$ $\therefore$ $\{H^+\} = 3.4 \cdot 10^{-4}$

(d) $\frac{\{NH_4^+\}\{OH^-\}}{\{NH_3\}} = K_b = 1.8 \cdot 10^{-5} = \frac{(0.15)\{OH^-\}}{(0.25)}$ $\{OH^-\} = 3.0 \cdot 10^{-5}$ $\therefore$ $\{H^+\} = 3.3 \cdot 10^{-1}$

(e) $\frac{\{N_2H_5^+\}\{OH^-\}}{\{N_2H_4\}} = K_b = 1.7 \cdot 10^{-6} = \frac{(0.50)\{OH^-\}}{(0.30)}$ $\{OH^-\} = 1.0 \cdot 10^{-6}$ $\therefore$ $\{H^+\} = 9.8 \cdot 10^{-9}$

15.35 pH = 11.40 $\{H^+\} = 4.0 \cdot 10^{-12}$ $\therefore$ $\{OH^-\} = 2.5 \cdot 10^{-3} \approx \{B^+\}$

$$\frac{\{OH^-\}\{B^+\}}{\{BOH\}} = K_b = \frac{(2.5 \cdot 10^{-3})^2}{(0.012-0.0025)} = 6.6 \cdot 10^{-4}$$

15.36 $HC_2H_3O_2 \rightleftharpoons H^+ + C_2H_3O_2^-$ $K_a = 1.8 \cdot 10^{-5} = \frac{H^+ \; C_2H_3O_2}{HC_2H_3O_2}$

The solution of this problem will require an extra column in the table to account for the addition of HCl. (This column will be headed 'ADD'.) In addition to the X which has been used for a change in concentration, a second unknown(Y) will be needed to represent the unknown concentration of HCl which existed immediately after that substance was added and before the change in concentration which accompanie attainment of equilibrium.

(THE TABLE AND THE COMPLETION OF THIS PROBLEM SOLUTION WILL BE FOUND ON THE FOLLOWING PAGE.)

(15.36 continued) LET Y = the instantaneous increase in $\{H^+\}$ (from HCl)

conc.	initial	added:	change	EQUILIBRIUM	assume:	$\therefore$
$HC_2H_3O_2$	0	0	+ X	X		X
H^+	10^{-7}	Y	− X	10^{-7}+Y−X	$Y >> 10^{-7}$	Y − X
$C_2H_3O_2^-$	1.0	0	− X	1.0 − X		1.0−X

let X = $\Delta\{HC_2H_3O_2\}$ during the "change" step pH = 4.74 $\therefore \{H^+\} = 1.8 \cdot 10^{-5}$

$$\frac{(1.8 \cdot 10^{-5})(1.0-X)}{X} = 1.8 \cdot 10^{-5} \quad \therefore X = 0.50$$

Since Y − X = $\{H^+\} = 1.8 \cdot 10^{-5}$

Y ≈ X = 0.50 M = molarity of HCl (added)

$$?\,gHCl = 0.500\,\cancel{ltr\,soln} \cdot \frac{0.50\,\cancel{mol\,HCl}}{1^*\,\cancel{ltr\,soln}} \cdot \frac{36.46\,g\,HCl}{1^*\,\cancel{mol\,HCl}} = \underline{\underline{9.1\,g\,HCl}}$$

15.37 $H_2C_6H_6O_6 \rightleftharpoons H^+ + HC_6H_6O_6^-$ $K_a = 7.9 \cdot 10^{-5} = \dfrac{\{H^+\}\{HC_6H_6O_6^-\}}{\{H_2C_6H_6O_6\}}$

(let vC = vitamin C)

$$?M(vC) = ?mol(vC) = 1^*\,\cancel{ltr\,soln} \cdot \frac{.500\,\cancel{g(vC)}}{.250\,\cancel{ltr\,soln}} \cdot \frac{1^*\,mol(vC)}{176\,\cancel{g(vC)}} = \underline{0.0114\,M(vC)}$$

let X = $\Delta\{H^+\} \approx \{H^+\}_{eq} \approx \{HC_6H_6O_6^-\}_{eq}$ $\therefore \dfrac{X^2}{(0.0114)} = 7.9 \cdot 10^{-5}$ $\underline{X^2 = 9.0 \cdot 10^{-7}}$

$\underline{X = 9.5 \cdot 10^{-4} = H^+}$ $\underline{\underline{pH = 3.02}}$

15.38 $?M(vC) = ?mol(vC) = 1^*\,\cancel{ltr\,soln} \cdot \dfrac{0.500\,\cancel{g(vC)}}{0.200\,\cancel{ltr\,soln}} \cdot \dfrac{1^*\,mol(vC)}{176\,\cancel{g(vC)}} = \underline{0.0142M}$

(15.38 continued) $K_{a1} = 7.9\cdot10^{-5} = \dfrac{\{H^+\}\{HC_6H_6O_6^-\}}{\{H_2C_6H_6O_6\}}$ let $X = \Delta\{H^+\}$

conc.	initial	added:	change	EQUILIBRIUM	assume:	∴
$H_2C_6H_6O_6$	0	0.0142	- X	0.0142-- X	X<<.01	0.0142
H^+	0.1	0	+ X	0.1 + X		0.1
$HC_6H_6O_6^-$	0	0	+ X	X		X

pH = 1.0
$\{H^+\} = 0.1$

$\dfrac{(0.1)\cdot X}{0.0142} = 7.9\cdot10^{-5}$ $X = 1.1\cdot10^{-5}$ (assumption valid)

fraction dissociated = X/0.0142 = $7.9\cdot10^{-4}$

15.39 (HBa = barbituric acid NaBa = sodium barbituate)

$HBa \rightleftharpoons H^+ + Ba^-$ $K_a = 1.0\cdot10^{-5} = \dfrac{\{H^+\}\{Ba^-\}}{\{HBa\}}$

?M(NaBa)= ?mol(NaBa)=1*~~ltr soln~~$\cdot\dfrac{0.050\text{ g(NaBa)}}{0.250\text{ ltr soln}}\cdot\dfrac{1\text{ mol(NaBa)}}{150\text{ g(NaBa)}}$ = $2.7\cdot10^{-4}$M

conc.	initial	added:	change	EQUILIBRIUM	Assume:	∴
HBa	0	0	+ X	X		X
H^+	0.1	0	- X	0.1 - X	X<<0.1	0.1
Ba^-	0	$2.7\cdot10^{-4}$	-X	$2.7\cdot10^{-4}$-X		$2.7\cdot10^{-4}$-X

$\dfrac{(0.1)(2.7\cdot10^{-4} - X)}{X} = 1.0\cdot10^{-5}$ $2.7\cdot10^{-5} - 0.1X = 1.0\cdot10^{-5}X$

$X = 2.7\cdot10^{-4} = \{HBa\}$ This result demonstrates that essentially all of the added sodium barbituate is converted to barbituric acid. (100%)

15.40 (HNic = nicotinic acid) $HNic \rightleftharpoons H^+ + Nic^-$

$K_a = 1.4 \cdot 10^{-5} = \frac{\{H^+\}\{Nic^-\}}{\{HNic\}} = \frac{(X)(X)}{(0.010-X)}$ let $X = \Delta\{H^+\}$
if we assume: $X << 0.010$ then:

$\frac{X^2}{(0.010)} = 1.4 \cdot 10^{-5}$ $X^2 = 1.4 \cdot 10^{-7}$ $X = 3.7 \cdot 10^{-4}$ (assumption valid)

$\{H^+\} = 3.7 \cdot 10^{-4}$ $pH = 3.43$

15.41 $pH = 2.5$ $H^+ = 3.2 \cdot 10^{-3}$ $HC_2H_3O_2 \rightleftharpoons H^+ + C_2H_3O_2^-$

$\{C_2H_3O_2^-\} \simeq \{H^+\} = 3.2 \cdot 10^{-3}$ since: $\frac{\{H^+\}\{C_2H_3O_2^-\}}{\{HC_2H_3O_2\}} = 1.8 \cdot 10^{-5}$ then:

$\frac{(3.2 \cdot 10^{-3})^2}{HC_2H_3O_2} = 1.8 \cdot 10^{-5}$ $\{HC_2H_3O_2\} = 0.57M$ (Since ionization is such a small fraction, the molarity of the acetic acid solution $\simeq$ the $\{HC_2H_3O_2\}_{EQ}$.)

15.42 $N_2H_4 + H_2O \rightleftharpoons N_2H_5^+ + OH^-$ $K_b = \frac{\{N_2H_5^+\}\{OH^-\}}{\{N_2H_4\}} = 1.7 \cdot 10^{-6}$

$pH = 10.64$ $\therefore$ $pOH = 3.36$ $\{OH^-\} = 4.4 \cdot 10^{-4}$ since: $\{N_2H_5^+\} \simeq \{OH^-\}$

then: $1.7 \cdot 10^{-6} = \frac{(4.4 \cdot 10^{-4})^2}{\{N_2H_4\}}$ $\{N_2H_4\} = 0.11M$ (see comment 15.41)

15.43 $HA \rightleftharpoons H^+ + A^-$ $\quad K_a = \dfrac{\{H^+\}\{A^-\}}{\{HA\}}$ $\quad pH = 4.55$

$\{H^+\} = 2.82 \cdot 10^{-5} \simeq \{A^-\}$ $\therefore$ $\dfrac{(2.82 \cdot 10^{-5})^2}{(0.010)} = K_b = 7.9 \cdot 10^{-8}$

15.44 $H_2C_6H_6O_6 \rightleftharpoons H^+ + HC_6H_6O_6^-$ $\quad K_{a1} = \dfrac{\{H^+\}\{HC_6H_6O_6^-\}}{\{H_2C_6H_6O_6\}} = 7.9 \cdot 10^{-5}$

$K_{a2} = \dfrac{\{H^+\}\{C_6H_6O_6^{2-}\}}{\{HC_6H_6O_6^-\}} = 1.6 \cdot 10^{-12}$ $\quad HC_6H_6O_6^- \rightleftharpoons H^+ + C_6H_6O_6^{2-}$

let $X = \Delta\{H^+\}$ due to the first ionization, $Y = \Delta\{H^+\}$ due to the second

conc.	initial	added:	chng 1	chng 2	EQUILIBRIUM	assume:	$\therefore$
H^+	10^{-7}	0	+ X	+ Y	10^{-7}+X+Y	$10^{-7} << X$	X
$H_2C_6H_6O_6$	0	0.050	− X	0	0.050 − X	Y<<X	0.050
$HC_6H_6O_6^-$	0	0	+ X	− Y	X − Y	X<<0.050	X
$C_6H_6O_6^{2-}$	0	0	0	+ Y	Y		Y

$\dfrac{X^2}{(0.050)} = 7.9 \cdot 10^{-5}$ $\quad X^2 = 3.95 \cdot 10^{-6}$ $\quad X = 2.0 \cdot 10^{-3}$ (1st & 3rd assumptions are valid) $\quad \dfrac{(X)(Y)}{X} = 1.6 \cdot 10^{-12}$ $\therefore$ $Y = 1.6 \cdot 10^{-12}$ (2nd assumption is valid) $\therefore$ $\{H^+\} = 2.0 \cdot 10^{-3}$ $\quad \{H_2C_6H_6O_6\} = 0.050$

$\{HC_6H_6O_6^-\} = 2.0 \cdot 10^{-3}$ $\quad \{C_6H_6O_6^{2-}\} = 1.6 \cdot 10^{-12}$

NOTE: The simplifying assumptions made the arithmetic easy, but the answers (to 2 s.f.) are the same as those which could be obtained by an exact algebraic solution (much more difficult!).

IT IS A SOUND IDEA TO BE LAZY IN A PRODUCTIVE WAY!

15.45 $H_3PO_4 \rightleftharpoons H^+ + H_2PO_4^-$ $H_2PO_4^- \rightleftharpoons H^+ + HPO_4^{2-}$ $HPO_4^{2-} \rightleftharpoons H^+ + PO_4^{3-}$

$$K_{a1} = \frac{\{H^+\}\{H_2PO_4^-\}}{\{H_3PO_4\}} = 7.5 \bullet 10^{-3} \qquad K_{a2} = \frac{\{H^+\}\{HPO_4^{2-}\}}{\{H_2PO_4^-\}} = 6.2 \bullet 10^{-8}$$

$$K_{a3} = \frac{\{H^+\}\{PO_4^{3-}\}}{\{HPO_4^{2-}\}} = 2.2 \bullet 10^{-12}$$

let $X = \Delta\{H^+\}$ (first ionization) let $Y = \Delta\{H^+\}$ (2nd) let $Z = \Delta\{H^+\}$ (3rd)

conc.	initial	chng 1	chng 2	chng 3	EQUILIBRIUM	assume:	∴
H^+	10^{-7}	+ X	+ Y	+ Z	10^{-7}+X+Y+Z	10^{-7}<<X	X
H_3PO_4	1.0	− X	0	0	1.0 − X	X<<1.0	1.0
$H_2PO_4^-$	0	+ X	− Y	0	X − Y	Y<<X	X
HPO_4^{2-}	0	0	+ Y	− Z	Y − Z	Z<<Y	Y
PO_4^{3-}	0	0	0	+ Z	Z		Z

$$\frac{X^2}{1.0} = 7.5 \bullet 10^{-3}$$ $X = 0.0866$ (1st & 2nd assumptions valid) $$\frac{X \bullet Y}{X} = 6.2 \bullet 10^{-8}$$ $Y = 6.2 \bullet 10^{-8}$ (3rd assumption valid) $$\frac{X \bullet Z}{Y} = 2.2 \bullet 10^{-12}$$

$$\frac{(0.0866) \bullet Z}{(6.2 \bullet 10^{-8})} = 2.2 \bullet 10^{-12}$$ $Z = 1.6 \bullet 10^{-18}$ (4th assumption valid)

$\{H^+\} = \{H_2PO_4^-\} = 8.7 \bullet 10^{-2}M$ $\{H_3PO_4\} = 1.0M$ $\{HPO_4^{2-}\} = 6.2 \bullet 10^{-8}M$

$\{PO_4^{3-}\} = 1.6 \bullet 10^{-18}M$

15.46 $H_2SeO_3 \rightleftharpoons HSeO_3^- + H^+$ $HSeO_3^- \rightleftharpoons SeO_3^{2-} + H^+$

(15.46 continued) $K_{a1} = \frac{\{H^+\}\{HSeO_3^-\}}{\{H_2SeO_3\}} = 3 \cdot 10^{-3}$ $K_{a2} = \frac{\{H^+\}\{SeO_3^{2-}\}}{\{HSeO_3^-\}} =$ $5 \cdot 10^{-8}$ let $X = \Delta\{H^+\}$ (1st ionization) let $Y = \Delta\{H^+\}$ (2nd)

conc.	initial	chng 1	chng 2	EQUILIBRIUM	ASSUME:	∴
H^+	10^{-7}	+ X	+ Y	10^{-7} + X + Y	$X >> 10^{-7}$	X
H_2SeO_3	0.50	- X	0	0.50 - X	X>>Y	0.50
$HSeO_4^-$	0	+ X	- Y	X - Y	X<<0.5	X
SeO_4^{2-}	0	0	+ Y	Y		Y

$\frac{X^2}{0.50} = 3 \cdot 10^{-3}$ X = 0.04 (1st assumption valid but 3rd is questionable since 0.04 is not negligible compared to 0.5) A second approximation however will show that to the 1 s.f. allowed (since the K_a values have only 1 s.f.) we obtain the same value of X as that calculated using the weak approximation! As demonstrated by the solutions to the last two problems: $Y = K_{a2} = 5 \cdot 10^{-8}$ (2nd assumption valid) $\{H^+\} = 0.04$ M

pH = 1.4 $\{H_2SeO_3\} = 0.50M$ $\{HSeO_3^-\} = 0.04M$ $\{SeO_3^{2-}\} = 5 \cdot 10^{-8}M$

15.47 $H_2CO_3 \rightleftharpoons H^+ + HCO_3^-$ $HCO_3^- \rightleftharpoons H^+ + CO_3^{2-}$

$\frac{\{H^+\}\{HCO_3^-\}}{\{H_2CO_3\}} = K_{a1}$ $\frac{\{H^+\}\{CO_3^{2-}\}}{\{HCO_3^-\}} = K_{a2}$ $K_{a1} = 4.3 \cdot 10^{-7}$ $K_{a2} = 5.6 \cdot 10^{-11}$

pH = 3.00 $\{H^+\} = 0.001$ assume: $\{H_2CO_3\}_{EQ} = 0.10M$ then:

$\frac{(0.001)\{HCO_3^-\}}{(0.10)} = 4.3 \cdot 10^{-7}$ $\{HCO_3^-\} = 4.3 \cdot 10^{-5}$ $\frac{(0.001)\{CO_3^-\}}{4.3 \cdot 10^{-5}} = 5.6 \cdot 10^{-11}$

(15.47 continued) $\{CO_3^{2-}\} = 2.4 \cdot 10^{-12} M$

15.48 $1.1 \cdot 10^{-21} = K_{a1} \cdot K_{a2} = \frac{\{H^+\}^2\{S^{2-}\}}{\{H_2S\}} = \frac{\{H^+\}^2(8.4 \cdot 10^{-15})}{(0.10)}$

$\{H^+\}^2 = 1.3 \cdot 10^{-8}$ $\{H^+\} = 1.1 \cdot 10^{-4} M$

15.49 $\frac{\{H^+\}^2\{S^{2-}\}}{\{H_2S\}} = 1.1 \cdot 10^{-21}$ pH = 4.60 $\{H^+\} = 2.5 \cdot 10^{-5}$

$\{S^{2-}\} = \frac{(1.1 \cdot 10^{-21})(0.10)}{(2.5 \cdot 10^{-5})^2} = 1.8 \cdot 10^{-13} M$

15.50 $\frac{\{H^+\}\{C_3H_5O_3^-\}}{\{HC_3H_5O_3\}} = 1.38 \cdot 10^{-4}$ if: pH = 4.25 then: $\{H^+\} = 5.6 \cdot 10^{-5}$

$\frac{\{C_3H_5O_3^-\}}{\{HC_3H_5O_3\}} = \frac{1.38 \cdot 10^{-4}}{5.6 \cdot 10^{-5}} = 2.46$ RATIO = 0.41 (reciprocal of value)

15.51 (a) $1.8 \cdot 10^{-5} = \frac{\{NH^+\}\{OH^-\}}{\{NH_3\}} = \frac{\cancel{(0.10)}\{OH^-\}}{\cancel{(0.10)}}$ $\therefore \{OH^-\} = 1.8 \cdot 10^{-5} M$

(15.51 continued) (b) $1.8 \bullet 10^{-5} = \frac{\{H^+\}\{C_2H_3O^-\}}{\{HC_2H_3O\}} = \frac{\{H^+\}(0.40)}{(0.20)}$ $\{H^+\}=9.0 \bullet 10^{-}$

(a) pH = 9.26 (b) pH = 5.05

(c) pH = 8.41 (d) CAREFUL! HCl is a strong acid. This system is not a buffer! $\{H^+\} = 0.20$ pH = 0.70

15.52 $\frac{\{H^+\}\{C_2H_3O^-\}}{\{HC_2H_3O\}} = 1.8 \bullet 10^{-5} = \frac{(7.1 \bullet 10^{-6})\{C_2H_3O^-\}}{(1.00)}$ $\{C_2H_3O^-\} \doteq 2.5M$

$?gNaC_2H_3O = 1.00\ \text{ltr soln} \bullet \frac{2.5^{\bullet}\ \text{mol}\ NaC_2H_3O}{1^{*}\ \text{ltr soln}} \bullet \frac{82.0\ g\ NaC_2H_3O}{1^{*}\ \text{mol}\ NaC_2H_3O} = 210\ g\ NaC_2H_3O$ (2 s.f.)

15.53 $\frac{\{NH_3\}}{\{OH^-\}\{NH_4^+\}} = \frac{1}{1.8 \bullet 10^{-5}}$ $\frac{\{NH_3\}}{\{NH_4^+\}} = \frac{10^{-4}}{1.8 \bullet 10^{-5}} = 5.6$ pH=10.0 $\{H^+\}=10^{-10}$ $\{OH^-\}= 10^{-4}$

15.54 $HC_2H_3O_2 \rightleftharpoons H^+ + C_2H_3O_2^-$ $K_a = \frac{\{H^+\}\{C_2H_3O_2^-\}}{\{HC_2H_3O_2\}} = 1.8 \bullet 10^{-5}$

pH = 3.0 $\{H^+\} = 0.0010$ let X = the instantaneous $\Delta\{H^+\}$ due to added HCl let Y = $\Delta\{HC_2H_3O_2\}$ during change step

(The table and the completion of this problem solution on next page.)

(15.54 continued)

conc.	initial	added:	change	EQUILIBRIUM	assume:	∴
H^+	10^{-7}	X	- Y	10^{-7} +X -Y	$X \gg 10^{-7}$	X - Y
$C_2H_3O_2^-$	0.010	0	- Y	0.010 - Y	$Y \gg 10^{-7}$	0.010-Y
$HC_2H_3O_2$	0.010	0	+ Y	0.010 + Y		0.010+Y

note: $\{H^+\}_{EQ} = 0.0010M = (X - Y)$ $\qquad \frac{(0.0010)(0.010-Y)}{(0.010+Y)} = 1.8 \bullet 10^{-5}$

$10^{-5} - 10^{-3}Y = 1.8 \bullet 10^{-7} + 1.8 \bullet 10^{-5}Y \qquad Y = 9.7 \bullet 10^{-3}$

X = 0.0010 + 0.0097 = 0.0107M As specified X = molarity of added HCl but the pH value allows only one s.f. so:

0.01 mol HCl per liter of solution

15.55 ΔpH = ? (a) The addition of 0.10 mol NaOH to a system which contains 0.50mol $HC_2H_3O_2$ and 0.50 mol $NaC_2H_3O_2$ yields: 0.40mol $HC_2H_3O_2$ and 0.60mol $NaC_2H_3O_2$. Since these amounts are contained in 0.5 liters, the concentrations are: $HC_2H_3O_2$ = 0.80M and $C_2H_3O_2^-$ = 1.20M

INITIAL: $\frac{\{H^+\}\{C_2H_3O_2^-\}}{\{HC_2H_3O_2\}} = 1.8 \bullet 10^{-5} = \frac{\{H^+\}(1.00)}{(1.00)} \qquad \{H^+\}_i = 1.8 \bullet 10^{-5}M \qquad pH_i = 4.74$

FINAL: $\frac{\{H^+\}(1.20)}{(0.80)} = 1.8 \bullet 10^{-5} \qquad \{H^+\}_f = 1.2 \bullet 10^{-5} \qquad pH_f = 4.92 \qquad \Delta pH = 0.18$

(b) (as 15.55a) ∴ pH_f = 5.11 $\qquad \Delta pH = 0.37$

(c) (as 15.55a) ∴ pH_f = 5.70 $\qquad \Delta pH = 0.59$ (In part c the pH_i is equal to the pH_f in part b of the problem.)

(15.55 continued) In part d the added NaOH consumes all of the $HC_2H_3O_2$ which was present in the original buffer. After the pH of the original buffer is calculated the pH of a 1.00M $NaC_2H_3O_2$ solution will be calculated and the ΔpH determined.

$$\frac{\{H^+\}(0.80)}{(0.20)} = 1.8 \cdot 10^{-5} \qquad \{H^+\}_i = 4.5 \cdot 10^{-6} \qquad pH_i = 5.35$$

FINAL: let X = Δ{OH⁻} due to hydrolysis of $NaC_2H_3O_2$ $\frac{\{OH^-\}\{HC_2H_3O_2\}}{\{C_2H_3O_2^-\}} = K_{hy}$

$$K_{hy} = K_w/K_a = 5.56 \cdot 10^{-10} \qquad \frac{X^2}{1.00} = 5.56 \cdot 10^{-10} \qquad X = 2.4 \cdot 10^{-5} = \{OH^-\}$$

$$pOH = 4.62 \qquad pH_f = 9.38 \qquad \Delta pH = 4.03$$

(e) The pH_i in part e equals the pH_f in part c. $pH_i = 5.70$

FINAL: let X = Δ{OH⁻} due to hydrolysis of $NaC_2H_3O_2$ After the addition of NaOH, $\{C_2H_3O_2^-\}$ = 1.00M and {OH⁻} = 0.10M. Substituting into the hydrolysis constant expression shown in part d yields:

$$\frac{(0.10 + X)(X)}{(1.00 - X)} = 5.56 \cdot 10^{-10}$$

If we assume that X<<0.10 then the expression simplifies to: $(0.10X)/(1.00) = K_{hy}$

and $X = 5.6 \cdot 10^{-9}$ $\therefore \{OH^-\}_{EQ} = 0.10M$ pOH = 1.00 pH = 13.00

ΔpH = 7.30

15.56 $HCHO_2 \rightleftharpoons H^+ + CHO_2^-$ $K_a = \frac{\{H^+\}\{CHO_2^-\}}{\{HCHO_2\}} = 1.8 \cdot 10^{-4}$

let X = Δ{$HCHO_2$} during change step

conc.	initial	added:	change	EQUILIBRIUM	assume:	∴
$HCHO_2$	0.45	0	+ X	0.45 + X	X≃0.1	0.55
H^+	10^{-7}	0.10	- X	0.10 - X		very small
CHO_2^-	0.55	0	- X	0.55 - X		0.45

(15.56 continued) $\frac{\{H^+\}(0.45)}{(0.55)} = 1.8 \cdot 10^{-4}$ $\{H^+\}_f = 2.2 \cdot 10^{-4} M$

BEFORE THE ADDITION OF HCl: $\frac{\{H^+\}(0.55)}{(0.45)} = 1.8 \cdot 10^{-4}$ $\{H^+\}_i = 1.5 \cdot 10^{-4} M$

$pH_f = 3.66$ $pH_i = 3.82$ $\Delta pH = -0.16$

15.57 ($pH_i = 3.82$ as calculated in 15.56) AFTER ADDITION OF NaOH:

$$\frac{\{H^+\}(0.55 + 0.20)}{(0.45 - 0.20)} = 1.8 \cdot 10^{-4} \qquad \{H^+\}_f = 6.0 \cdot 10^{-5} M$$

$pH_f = 4.22$ $\Delta pH = 0.40$

15.58 $C_2H_3O_2^- + H_2O \rightleftharpoons HC_2H_3O_2 + OH^-$ $K_{hy} = \frac{K_w}{K_a} = \frac{\{OH^-\}\{HC_2H_3O_2\}}{\{C_2H_3O_2^-\}}$

$K_{hy} = 5.56 \cdot 10^{-10}$ $\{OH^-\} \simeq \{HC_2H_3O_2\}$ $\frac{\{OH^-\}^2}{(1.0 \cdot 10^{-3})} = 5.56 \cdot 10^{-10}$

$\{OH^-\} = 7.5 \cdot 10^{-7} M$ $pOH = 6.12$ $pH = 7.88$ (a)

b) $\{NH_3\} \simeq \{H^+\} = 8.4 \cdot 10^{-6} M$ $pH = 5.08$

c) $\{HCO_3^-\} \simeq \{OH^-\} = 4.2 \cdot 10^{-3} M$ $pH = 11.63$

d) $\{HCN\} \simeq \{OH^-\} = 1.4 \cdot 10^{-3} M$ $pH = 11.15$

e) $\{NH_2OH\} \simeq \{H^+\} = 4.3 \cdot 10^{-4}$ $pH = 3.37$

15.59 (Pyr) = pyridine $(Pyr)H^+$ = pyridinium ion

$$(Pyr)H^+ + H_2O \rightleftharpoons H^+ + (Pyr) \qquad K_{hy} = \frac{K_w}{K_b} = \frac{\{H^+\}\{(Pyr)\}}{\{(Pyr)H^+\}}$$

$K_{hy} = 5.88 \bullet 10^{-6}$ let $X = \Delta\{H^+\}$ during change step

conc.	initial	added:	change	EQUILIBRIUM	assume:	∴
H^+	10^{-7}	0	+ X	10^{-7} + X	$X >> 10^{-7}$	X
(Pyr)	0	0	+ X	X	$0.10 >> X$	X
$(Pyr)H^+$	0.10	0	- X	0.10 - X		0.10

$$\frac{X^2}{0.10} = 5.88 \bullet 10^{-6} \qquad X = 7.67 \bullet 10^{-4} \qquad \%\text{hydrolysis} = \frac{7.67 \bullet 10^{-4}}{(0.10)} \bullet 100 = 0.77\%$$

15.60 HA = weak acid NaA = sodium salt $HA \rightleftharpoons H^+ + A^-$

$K_a = \frac{\{H^+\}\{A^-\}}{\{HA\}}$ pH = 9.35 $\{H^+\} = 4.47 \bullet 10^{-10}$ $\{A^-\} \simeq 0.10M$ $\{HA\} \simeq \{OH^-\}$
$2.24 \bullet 10^{-5}M$ $\therefore$ $\frac{(4.47 \bullet 10^{-10})(0.10)}{(2.24 \bullet 10^{-5})} = 2.0 \bullet 10^{-6} = K_a$

15.61 $OCl^- + H_2O \rightleftharpoons HOCl + OH^-$ $K_{hy} = \frac{10^{-14}}{3.1 \bullet 10^{-8}} = \frac{\{HOCl\}\{OH^-\}}{\{OCl^-\}}$

let $X = \Delta\{OH^-\}$ during the change step

$\frac{(X)(10^{-7}+X)}{(0.67-X)} = 3.23 \bullet 10^{-7}$ assume: $0.67 >> X >> 10^{-7}$ then: $X^2 = 2.16 \bullet 10^{-7}$

$X = 4.7 \bullet 10^{-4} = \{OH^-\}$ pOH=3.33 pH = 10.67

15.62 H(Ver) = veronal   Na(Ver) = sodium salt

$$?M\ Na(Ver) = ?mol\ Na(Ver) = 1^{*}\cancel{ltr\ soln}\cdot\frac{10\cdot10^{-3}\cancel{g\,Na(Ver)}}{0.25\cancel{ltr\ soln}}\cdot\frac{1^{*}mol\ Na(Ver)}{206\cancel{g\ Na(Ver)}}$$

$= 1.9\cdot10^{-4}M$ $(Ver)^- + H_2O \rightleftharpoons H(Ver) + OH^-$ $\frac{\{H(Ver)\}\{OH^-\}}{\{(Ver)^-\}} =$

$K_{hy} = K_w/K_a = 2.7\cdot10^{-7} = \frac{\{OH^-\}^2}{1.9\cdot10^{-4}}$ $\{OH^-\}^2 = 5.13\cdot10^{-11}$ $\{OH^-\}=7.2\cdot10^{-6}$

pOH = 5.14   pH = 8.86

15.63 H(Barb) = barbituric acid   Na(Barb) = sodium salt

$K_{hy} = \frac{K_w}{K_a} = \frac{\{H(Barb)\}\{OH^-\}}{\{(Barb)^-\}}$ let $X = \Delta\{OH^-\}$ during the change step

$$K_{hy} = \frac{10^{-14}}{1.0\cdot10^{-5}} = 1.0\cdot10^{-9} = \frac{X(10^{-7} + X)}{1.0\cdot10^{-3}-X}$$

assume: $1.0\cdot10^{-3} >> X >> 10^{-7}$ then: $\frac{X^2}{1.0\cdot10^{-3}} = 1.0\cdot10^{-9}$ $X^2 = 1.0\cdot10^{-12}$

$X = 1.0\cdot10^{-6} \simeq \{H(Barb)\}$

15.64 H(vC) = ascorbic acid   Na(vC) = sodium salt

$K_{hy} = \frac{K_w}{K_a} = \frac{10^{-14}}{1.6\cdot10^{-12}} = 6.25\cdot10^{-3} = \frac{\{H(vC)\}\{OH^-\}}{\{(vC)^-\}}$ let $X = \Delta\{OH^-\}$ assume: $0.2 >> X >> 10^{-7}$

then: $X^2/0.20 = 6.25\cdot10^{-3}$ X = 0.035   pOH = 1.45   pH = 12.55 (cont.)

(15.64 continued) Although the assumption that X 10^{-7} was valid since X = 0.035, the assumption that 0.20 X was questionable since 0.035 is 17.5% of 0.20, not negligible. Successive approximations, or the solution of the quadratic equation however, only refine the value of X (to 2 s.f.) to 0.032 which yields a pH value of 12.51 not a very significant correction. Even with our shaky assumption we achieved a satisfactory final answer!

15.65 $PO_4^{3-} + H_2O \rightleftharpoons HPO_4^{2-} + OH^-$ $K_{hy} = \frac{K_w}{K_a} = \frac{\{OH^-\}\{HPO_4^{2-}\}}{\{PO_4^{3-}\}} =$

$\frac{10^{-14}}{2.2\cdot10^{-12}} = 4.55\cdot10^{-3}$ let X = $\Delta\{OH^-\}$ during change step $\{HPO_4^{2-}\}\simeq\{OH^-\}$

$4.55\cdot10^{-3} = \frac{(10^{-7}+X)X}{(0.50-X)}$ assume: $0.50 >> X >> 10^{-7}$ then: $X^2/0.50 = 4.55\cdot10^{-3}$ $\therefore X^2 = 2.28\cdot10^{-3}$ X = 0.0477 Although the assumption that X>>10^{-7} was valid, X (0.0477) is not negligible compared to 0.50. We can refine the initial value calculated for X by the method of successive approximations. When we assumed that X was very small compared to 0.50 we approximated the denominator of the mass action expression as 0.50. A much more precise value for that term can be obtained by using the initial value calculated for X. The denominator then becomes (0.50-0.0477) = 0.45 and the mass action expression equals X² divided by 0.45. X² then equals 2.06$\cdot$10^{-3} and a more refined value for X is calculated to be 0.0454 as a second approximation. A third approximation can be made using this new value of X to further refine the value of the denominator and then calculate a third approximation for X but it is not likely that the third approximation will change the value of X expressed to 2 s.f. (As a matter of fact X_3 = 0.045 as we guessed

Since X $\simeq\{OH^-\}$ pOH = 1.35 pH = 12.65 NOTE: Although HPO_4^{2-} can hydrolyze to $H_2PO_4^-$ and that species to H_3PO_4 these processes will not make a significant contribution to the equilibrium concentration of OH^- (In the same way, the equilibrium concentration of H^+ in a solution of a weak polyprotic acid can be determined by considering only the first ionization stage.)

15.66 $CN^- + H_2O \rightleftharpoons HCN + OH^-$ $\quad K_{hy} = \frac{K_w}{K_a} = \frac{\{OH^-\}\{HCN\}}{\{CN^-\}} = \frac{10^{-14}}{4.9\cdot10^{-10}}$

$= 2.04\cdot10^{-5}$ let $X = \Delta\{OH^-\}$ during change step THEN: the mass action

expression $= \frac{(10^{-7}+X)\cdot X}{(10^{-3}-X)}$ assume: $10^{-3} >> X >> 10^{-7}$ which allows the mass action expression to be simplified to:

$\frac{X^2}{(0.0010)} = 2.04\cdot10^{-5}$ $\quad X^2 = 2.04\cdot10^{-8}$ $\quad X = 1.43\cdot10^{-4}$ (As in 15.65 one of our assumptions was shaky. A second approximation (as in 15.65) yields $X_2 = 1.32\cdot10^{-4} \approx \{OH^-\}$. $\therefore$ pOH = 3.88 pH = 10.12

NOTE: To 2 s.f. the solution of the quadratic equation gives the same value for X.

15.67 $HNO_3 + KOH \rightleftharpoons H_2O + KNO_3$ At the equivalence point 15.0ml of 0.0200M HNO_3 has been titrated with 30.0ml of 0.0100M KOH to yield 45.0ml of 0.00667M KNO_3. Since this is a solution of the salt of a strong acid and a strong base the solution has a pH = 7.00.(a)

(b)(as indicated in part (a)) 30.0ml KOH

(c) ?mol H^+(XS) = 15.5ml(HNO_3)$\cdot$(0.0200mol(HNO_3)/10^3*ml(HNO_3) - 15.0ml (KOH)$\cdot$(0.0100mol(KOH)/10^3*ml(KOH) = $2.00\cdot10^{-4}$mol H^+(XS) SINCE THE FINAL VOLUME = 45.0ml $\therefore$ $\{H^+\} = 2.00\cdot10^{-4}/0.025 = 8.0\cdot10^{-3}$M pH = 2.10

(d) ?mol OH^-(XS) = 35.0~~ml(KOH)~~$\cdot\frac{0.0100\text{mol(KOH)}}{10^3\text{~~*ml(KOH)~~}}$ - 15.0~~ml(HNO_3)~~$\cdot\frac{0.0200\text{mol}}{10^3\text{~~*ml~~}}$ ~~(HNO_3)~~

$= 5.0\cdot10^{-5}$mol OH^-(XS) $\quad \{OH^-\} = 5.0\cdot10^{-5}/0.050 = 1.0\cdot10^{-3}$

pOH = 3.00 pH = 11.00

15.68 H(Barb) = barbituric acid Na(Barb) = sodium salt

At the equivalence point 25.0ml of acid have been titrated with 12.5ml of base producing $(0.025 \bullet 0.010) = 2.5 \bullet 10^{-4}$mol Na(Barb). Since the salt is contained in 37.5ml of solution: $6.67 \bullet 10^{-3}$M Na(Barb)

Since Na(Barb) is the salt of a weak acid and a strong base it will undergo hydrolysis: $(Barb)^- + H_2O \rightleftharpoons H(Barb) + OH^-$

let $X = \Delta\{OH^-\}$ during the change step: $\frac{\{H(Barb)\}\{OH^-\}}{\{(Barb)^-\}} = \frac{X \bullet (10^{-7}+X)}{(0.00667-X)} = K_{hy}$

assume: $0.00667 >> X >> 10^{-7}$ then:

$$\frac{X^2}{(0.00667)} = \frac{10^{-14}}{1.0 \bullet 10^{-5}} = 1.0 \bullet 10^{-9} \qquad X^2 = 6.67 \bullet 10^{-12} \qquad X = 2.6 \bullet 10^{-6}$$

(both assumptions are satisfactory)

$\{OH^-\} = 2.6 \bullet 10^{-6}$M pOH = 5.59 pH = 8.41

15.69 (a) After 5.0ml of base have been added the total volume of solution = 55.0ml. All of the concentrations in the table below have been adjusted to this value* let $X = \Delta\{H^+\}$ during change

conc.	initial	added:	change	EQUILIBRIUM	assume:	∴
HF	0.182*	-.0091*†	- X	0.173* - X	0.17>>X	0.173*
H^+	10^{-7}	0	+ X	10^{-7} + X	$X >> 10^{-7}$	X
F^-	0	+.0091*†	+ X	0.0091 + X		$9.1 \bullet 10^{-3}$+X

†The added base was adjusted for a dilution of 5 to 55ml and was then considered to reduce the concentration of HF and increase that of F^- by the adjusted concentration of added base. $HF \rightleftharpoons H^+ + F^-$

$$\frac{\{H^+\}\{F^-\}}{\{HF\}} = K_a = 6.5 \bullet 10^{-4} = \frac{X \bullet (9.1 \bullet 10^{-3}+X)}{(0.173)} \qquad X^2 + 9.1 \bullet 10^{-3}X = 1.12 \bullet 10^{-4}$$

QUADRATIC ROOTS: $6.97 \bullet 10^{-3}$, $-1.61 \bullet 10^{-2}$ ∴ $\{H^+\} = 6.97 \bullet 10^{-3}$M pH = 2.16

(15.69 continued) (b) Exactly the same technique will be employed as that used in part (a). In part (b) half neutralization would involve the addition of 50.0ml 0.100M NaOH to50.0ml of 0.200M HF giving a final volume of 100.0ml solution. The concentrations in the table below have all been adjusted to one half of the initial values.

conc.	initial	added:	change	EQUILIBRIUM	assume:	∴
HF	0.100*	-0.050†	- X	0.050 - X	0.050 X	0.050
H^+	10^{-7}	0	+ X	10^{-7} + X	X 10^{-7}	X
F^-	0	+0.050†	+ X	0.050 + X		0.050

† As in part (a) the added base was considered to reduce the concentration of HF and increase the concentration of F^-.

$$\frac{X\bullet(0.050)}{(0.050)} = 6.5\bullet10^{-4} \qquad X = 6.5\bullet10^{-4} \text{ (both assumptions valid)}$$

$\{H^+\} = 6.5\bullet10^{-4}$ pH = 3.19 (b)

(c) In part (c) the equivalence point of the titration is reached when 100.0ml of base have been added to the 50.0ml of HF. At this point we will consider that the solution contains (0.200 x 0.050)mol of NaF in a total volume of 150.0ml. Since NaF is the salt of a weak acid and a strong base it will undergo hydrolysis: $F^- + H_2O \rightleftharpoons HF + OH^-$

$$\frac{\{HF\}\{OH^-\}}{\{F^-\}} = K_{hy} = \frac{K_W}{K_a} = \frac{10^{-14}}{6.5\bullet10^{-4}} = 1.54\bullet10^{-11}$$ let X = $\Delta\{OH^-\}$ (change)

$$\frac{X\bullet(10^{-7}+X)}{(6.67\bullet10^{-2}-X} = K_{hy}$$ assume: $0.0667 >> X >> 10^{-7}$ then the mass action expression simplifies to: $X^2/6.67\bullet10^{-2} = 1.54\bullet10^{-11}$

$X^2 = 1.027\bullet10^{-12}$ $X = 1.0\bullet10^{-6}$ $\{OH^-\}$ pOH = 5.99 pH = 8.01

One of the assumptions made in part (c) is shaky! 10^{-7} is 10% of 10^{-6} and certainly not negligible compared to 10^{-6}. If we add the OH^- ion produced by the dissociation of water the $\{OH^-\} = 1.1\bullet10^{-6}$, the pOH = 5.96 and the pH = 8.04 which is not a very significant change.

15.70 H(Bz) = benzoic acid Na(Bz) = sodium salt

hydrolysis: $(Bz)^- + H_2O \rightleftharpoons H(Bz) + OH^-$ $\dfrac{\{H(Bz)\}\{OH^-\}}{\{(Bz)^-\}}$

$= \dfrac{K_w}{K_a} = 10^{-14}/6.5\bullet10^{-5} = 1.54\bullet10^{-10} = K_{hy}$ let $X = \Delta\{OH^-\}$ (change step)

conc.	initial	change	EQUILIBRIUM	assume:	∴
$(Bz)^-$	0.020	- X	0.020 - X	0.02>>X	0.020
H(Bz)	0	+ X	X		X
OH^-	10^{-7}	+ X	10^{-7} + X	X>>10^{-7}	X

$\dfrac{X^2}{0.020} = 1.54\bullet10^{-10}$ $X^2 = 3.08\bullet10^{-12}$ $X = 1.75\bullet10^{-6} \simeq \{OH^-\}$

pOH = 5.76 pH = 8.24 (assumptions O.K.

15.71 $HC_4H_7O_2 \rightleftharpoons H^+ + C_4H_7O_2^-$ $K_a = 1.5\bullet10^{-5} = \dfrac{\{H^+\}\{C_4H_7O_2^-\}}{\{HC_4H_7O_2\}}$

for 0.10M acid: let $X = \Delta\{H^+\}$ change assume: $0.1 >> X >> 10^{-7}$ then:

$\dfrac{X^2}{0.10} = 1.5\bullet10^{-5}$ $X^2 = 1.5\bullet10^{-6}$ $X = 1.22\bullet10^{-3}$ (both assumptions are valid)

$pH_1 = 2.91$ (initial value before the addition of any base)

Since solid base is being added to 100ml solution, the effective concentrations of base (before reaction) would be 10 x the molar quantities. After each addition of base: $\{acid\}_f = \{acid\}_i - \{base\}$ {conjugate base} = {base added}

0.0010 mol base added: $\{acid\}_2 = 0.10 - 0.010 = 0.09$ M $\{C_4H_7O_2^-\} = 0.010$M

$\dfrac{\{H^+\}(0.010)}{(0.090)} = 1.5\bullet10^{-5}$ $\{H^+\}_2 = 1.35\bullet10^{-4}$ $pH_2 = 3.87$

0.0050 mol base added: $\{acid\}_3 = (0.10 - 0.050) = 0.050$M $\{C_4H_7O_2^-\} = 0.050$M

(15.71 continued) $\frac{\{H^+\}\ (0.050)}{(0.050)} = 1.5\cdot10^{-5}$ $\{H^+\}_3 = 1.5\cdot10^{-5}$ $pH_3 = 4.82$

0.0090 mol base added: $\{acid\}_4 = (0.10 - 0.090) = 0.010M$ $\{C_4H_7O_2^-\} = 0.090M$

$\frac{\{H^+\}(0.090)}{(0.010)} = 1.5\cdot10^{-5}$ $\{H^+\}_4 = 1.67\cdot10^{-6}M$ $pH_4 = 5.78$

0.010 mol base added: $\{acid\}_5 = (0.10 - 0.10) = 0M$ $\{C_4H_7O_2^-\} = 0.10M$

This is the equivalence point. Since $NaC_4H_7O_2$ is the salt of a weak acid and a strong base it will undergo hydrolysis.

$\frac{\{HC_4H_7O_2\}\{OH^-\}}{\{C_4H_7O_2^-\}} = K_{hy} = \frac{K_w}{K_a} = \frac{10^{-14}}{1.5\cdot10^{-5}} = 6.67\cdot10^{-10}$ let $X = \Delta\{OH^-\}$ assume: $0.1 >> X >> 10^{-7}$

then: $\frac{X^2}{0.10} = 6.67\cdot10^{-10}$ $X^2 = 6.67\cdot10^{-11}$ $X = 8.16\cdot10^{-6}$ (assumptions O.K.)

$pOH = 5.09$ $pH_5 = 8.91$

0.011 mol base added: $\{OH^-\} \simeq 0.010M$ $pOH = 2.00$ $pH = 12.00$

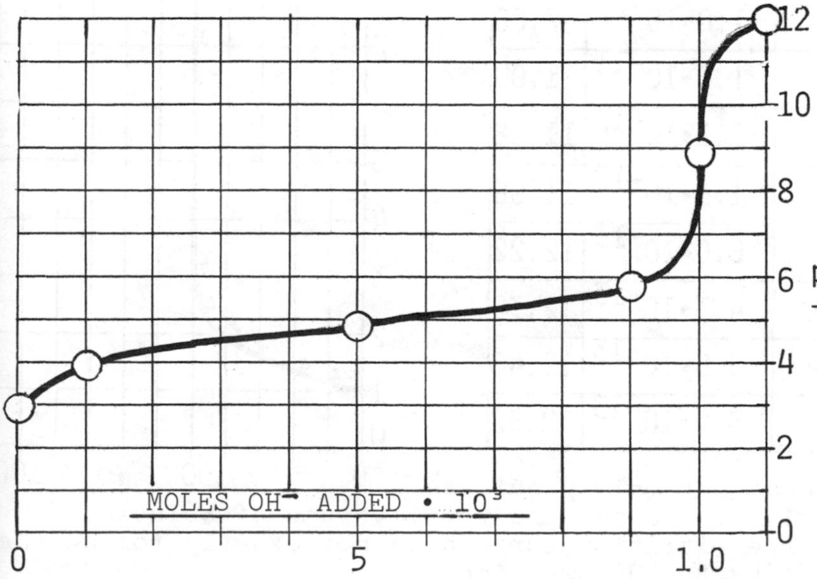

8.91 = pH @ equivalence point

INDICATOR: thymol blue OR phenolphthalein

15.72 $?\text{mol } H^+(\text{initial}) = 50.0\text{ml acid} \bullet \frac{0.10\text{mol } H^+}{10^3\text{*ml acid}} = \underline{5.0 \bullet 10^{-3}\text{mol } H^+}$

if mol H^+ > mol OH^- ∴ $\underline{\text{mol } H^+(XS) = \text{mol } H^+ - \text{mol } OH^-}$

if mol OH^- > mol H^+ ∴ $\underline{\text{mol } OH^-(XS) = \text{mol } OH^- - \text{mol } H^+}$

$\{H^+\}$ = mol H^+/final total volume in liters

$\{OH^-\}$ = mol OH^-/final total volume in liters

NOTE: IN THE TABLE BELOW THE XS H^+ (OR OH^-) AS MOLES, MUST BE DIVIDED BY THE CURRENT TOTAL VOLUME IN LITERS. (M)

point	ml base added:	mol base	M H^+	pH
1	0.00	0.00	0.10	1.00
2	10.00	0.0010	0.067	1.18
3	20.00	0.0020	0.043	1.37
4	30.00	0.0030	0.025	1.60
5	40.00	0.0040	0.011	1.95
6	45.00	0.0045	0.0053	2.28
7	49.00	0.0049	0.0010	3.00
8	50.00	0.0050	$1.0 \bullet 10^{-7}$	7.00
9	51.00	0.0051	$1.0 \bullet 10^{-11}$	11.00
10	55.00	0.0055	$2.1 \bullet 10^{-12}$	11.68
11	60.00	0.0060	$1.1 \bullet 10^{-12}$	11.96
12	70.00	0.0070	$6.0 \bullet 10^{-13}$	12.22
13	80.00	0.0080	$4.3 \bullet 10^{-13}$	12.36
14	90.00	0.0090	$3.5 \bullet 10^{-13}$	12.46
15	100.00	0.0100	$3.0 \bullet 10^{-13}$	12.52

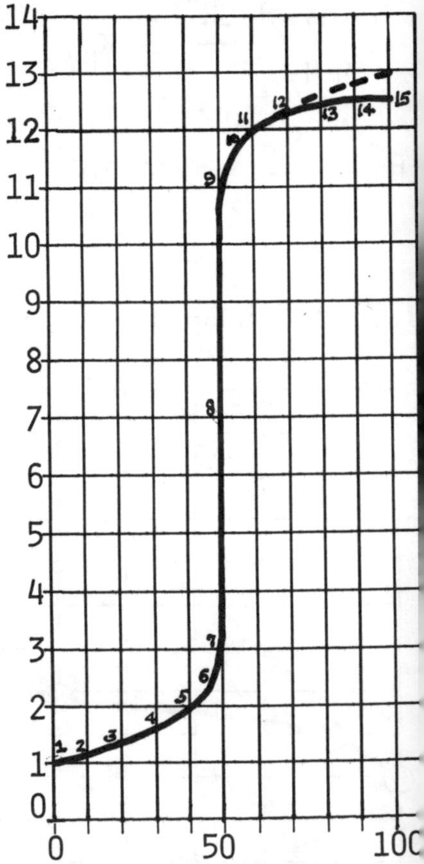

NOTE: Each addition of base not only provides additional OH^- ion but also increases the total volume of solution. This is the reason that the titration curve does not exactly follow the symetrical path indicated by the dashed line.

15.73 (a) HCN $pK_a = 9.31$ If we assume that the solution at the equivalence point is approximately 0.1M NaCN, the pH of the solution can be estimated by employing the Henderson-Hasselbalch equation (p481):

$$pH = pK_a + \log \frac{\{CN^-\}}{\{HCN\}}$$

If we assume that the concentration of CN^- is 100x greater than the HCN THE LOG OF 100 = 2 AND pH = $pK_a + 2 = 9.31 + 2 = 11.31$. The table shows that the range of alizarin yellow is from pH 10.1 to 12.0 so that indicator would be suitable.

(b) (for a base) $$pOH = pK_b + \log \frac{\{C_6H_5NH_3^+\}}{\{C_6H_5NH_2\}}$$

If we assume that the concentration of the ion is 100x greater than the concentration of the free base, $pOH = pK_b + 2 = 11.42$. ($pH \approx 2.6$)
The table shows that the range of thymol blue is from pH 1.2 to 2.8 so that indicator would be suitable. Although this method of estimating the pH at the equivalence point gives only an approximate value it is quite satisfactory for making a choice of an appropriate indicator.

15.74 (using the H-H equation) $$pH = pK_a + \log \frac{\{In^-\}}{\{HIn\}}$$

When the $\{In^-\} = \{HIn\}$ the log 1 = 0 and $pH = pK_a = 5$. A solution having a pH = 7 must have a log term = 2 $\therefore$ $\{In^-\} = 100 \times \{HIn\}$ so the solution would be green.

15.75 If we use the same approach for this problem as we have for similar problems in this chapter including: LET $X = \Delta\{H^+\}$ AND ASSUME: $0.0010 \gg X \gg 10^{-7}$ we calculate a value for $X = 1.3 \cdot 10^{-4}$. Since this value is certainly not negligible compared to 0.0010 one of our assumptions is shaky and our calculation of $\{H^+\} = 1.3 \cdot 10^{-4}$ is uncertain.

(15.75 continued) If we discard our simplifying assumption and solve the resulting quadratic equation we obtain an X valu of $1.25 \cdot 10^{-4}$ which to 2 s.f. yields EXACTLY THE SAME VALUE as our 'shak assumption. SHOW COURAGE! Make simplifying assumptions. Even when an assumption is shaky the result calculated by the use of the assumpti is often satisfactory to 2 s.f. Remember you can always do a successiv approximation to refine the orriginal answer.

15.76 $HCO_2H \rightleftharpoons H^+ + CO_2H^-$ $K_a = 1.8 \cdot 10^{-4} = \frac{\{H^+\}\{CO_2H^-\}}{\{HCO_2H\}}$

let X = $\Delta\{H^+\}$ during change step

conc.	initial	change	EQUILIBRIUM	assume	∴
HCO_2H	0.010	- X	0.010 - X	0.010>>X	0.010
H^+	10^{-7}	+ X	10^{-7} + X	X>>10^{-7}	X
CO_2H^-	0	+ X	X		X

$$\frac{X^2}{0.010} = 1.8 \cdot 10^{-4} \qquad X^2 = 1.8 \cdot 10^{-6} \qquad X = 1.34 \cdot 10^{-3}$$

This is one additional example: SHAKY ASSUMPTION→ANS O

∴ $\{H^+\} = 1.3 \cdot 10^{-3}M$ $\{HCO_2H\} = 0.010 - 1.34 \cdot 10^{-3} = 0.0087M$

$\{CO_2H^-\} = 1.3 \cdot 10^{-3}M$

15.77 This problem is more complex than most of those we have solved in this chapter because the HCO_3^- ion participates in two separ ate equilibria: $HCO_3^- \rightleftharpoons H^+ + CO_3^{2-}$ (ionization) AND (hydrolysis $HCO_3^- + H_2O \rightleftharpoons H_2CO_3 + OH^-$ IMPORTANT: ALTHOUGH BOTH EQUILIB RIA CONSUME HCO_3^- THERE IS ONLY ONE CONCENTRATION OF HCO_3^- WHICH MUST HAV A VALUE WHICH SATISFIES BOTH EQUILIBRIA.

(15.77 continued) $K_{a2} = 5.6\cdot10^{-11} = \frac{\{H^+\}\{CO_3^{2-}\}}{\{HCO_3^-\}}$ $\frac{\{H_2CO_3\}\{OH^-\}}{\{HCO_3^-\}} = K_{hy} =$

$\frac{K_W}{K_{a1}} = \frac{10^{-14}}{4.3\cdot10^{-7}} = 2.33\cdot10^{-8}$ Since the $\{HCO_3^-\}$ is the same in both equilibria, and $K_{hy} > K_{a2}$ we can predict that $\{OH^-\} > \{H^+\}$.

let X = $-\Delta\{HCO_3^-\}$ due to ionization and Y = $-\Delta\{HCO_3^-\}$ due to hydrolysis

conc.	initial	chng 1	chng 2	EQUILIBRIUM	assume	∴
HCO_3^-	0.50	- X	- Y	0.50-X-Y	0.50>>Y	0.50
H^+	10^{-7}	+ X	- X*	$<10^{-7}$	Y≈X	$10^{-14}/\{OH^-\}$
CO_3^{2-}	0	+ X	0	X		X
OH^-	10^{-7}	0	v.s.	$>10^{-7}$		$\{OH^-\}$
H_2CO_3	0	0	+ Y	Y		Y

* H^+ and OH^- combine to produce H_2O so that the $\{H^+\}\cdot\{OH^-\} = 10^{-14}$. ∴ the OH^- ion produced by hydrolysis consumes all of the H^+ produced by the ionization step (actually, some of the original 10^{-7} concentration of H^+ is also consumed so that the $\{H^+\}_{EQ} < 10^{-7}$) with an XS of OH^- remaining so that $\{OH^-\}_{EQ} > 10^{-7}$.

† Although we previously expressed our conviction that Y>X, it is also true that Y and X are expected to be close in value with only a small difference.

$$\frac{(10^{-14})/\{OH^-\}\cdot X}{0.50} = 5.6\cdot10^{-11} \qquad X = \frac{0.50\cdot5.6\cdot10^{-11}\cdot\{OH^-\}}{10^{-14}} = \underline{2.8\cdot10^{3}\{OH^-\}}$$

$$\frac{Y\cdot\{OH^-\}}{0.50} = 2.33\cdot10^{-8} \qquad Y = \frac{1.17\cdot10^{-8}}{\{OH^-\}}$$

Our assumption that Y≈X allows setting the factor which equals X in the first equation equal to the factor which equals Y in the second equation: $2.8\cdot10^{3}\{OH^-\} =$

$$1.17\cdot10^{-8}/\{OH^-\} \qquad \{OH^-\}^2 = 4.18\cdot10^{-12} \qquad \underline{\{OH^-\} = 2.04\cdot10^{-6}} \quad \therefore \underline{\underline{pH = 8.31}}$$

NOTE: From the calculated value of $\{OH^-\}$ substituted into the factor which equals X we can calculate $X = 5.7\cdot10^{-3}$.
Since $\{OH^-\} = 10^{-7} + Y - X$ then $Y = 2.0\cdot10^{-6} + 5.7\cdot10^{-3} - 10^{-7} = 5.7\cdot10^{-3}$ which demonstrates the validity of both assumptions.

(15.77 continued) PART 2: Δ(pH) with addition of HCl = ?

As a first step, we will refine the values calculated in part 1:

$\{HCO_3^-\}_{EQ} = 0.50 - X - Y = 0.50 - 5.6\cdot10^{-3} - 5.6\cdot10^{-3} = \underline{0.489M}$ ← (This is a more precise value than our approximation of 0.50M.)

Substitution of this value in the equations used in part 1 yields the same value for $\{OH^-\}$ ($2.04\cdot10^{-6}$) but $X = 5.57\cdot10^{-3} = Y = \{CO_3^{2-}\}\ \{H_2CO_3\}$

There are three species in the system which can react with the added H^+. In order of their base strengths these species are: OH^-, CO_3^{2-} and HCO_3^-. Since all of these species are involved in equilibria, their concentrations can not reduce to zero, but since the concentration of added H^+ is much larger than the concentrations of OH^- and CO_3^{2-}, we can assume that about 9/10 of each will be consumed and almost all of the remaining H^+ will react with HCO_3^-. This is detailed in the table below:

conc.	initial	added:	change	EQUILIBRIUM
OH^-	$2.04\cdot10^{-6}$		$-1.8\cdot10^{-6}$	$\{OH^-\}$
CO_3^{2-}	$5.57\cdot10^{-3}$		$-5.0\cdot10^{-3}$	$\{CO_3^{2-}\}$
HCO_3^-	0.489		-0.040†	0.449
H^+	$4.90\cdot10^{-9}$	0.05	-0.05®	$10^{-14}/\{OH^-\}$
H_2CO_3	$5.57\cdot10^{-3}$		+0.045*	$5.06\cdot10^{-2}$

The change is only a guess, $\{OH^-\}_{EQ} \simeq 10^{-7}$ $\{CO_3^{2-}\}_{EQ} \simeq 5\cdot10^{-4}$

†Although 0.045M of HCO_3^- was consumed, 0.005M was produced by the reaction of H^+ with CO_3^{2-}, so the <u>net</u> change = -0.040M.

®Almost all of the added H^+ will be consumed by reaction with the three bases present in the system. The equilibrium concentration of H^+ will be about 10^{-7}.

*(see †) the 0.045M of HCO_3^- which reacted with H^+ increased the $\{H_2CO_3\}$ by that amount.

USING:
$$\frac{\{H^+\}\{HCO_3^-\}}{\{H_2CO_3\}} = 4.3\cdot10^{-7} = \frac{\{H^+\}\cdot(0.449)}{(5.06\cdot10^{-2})} \qquad \underline{\{H^+\} = 4.8\cdot10^{-8}}$$

$$\underline{pH = 7.32} \qquad \therefore \ \underline{\underline{\Delta(pH) = -0.99}}$$

15.78 pH = 4.25 $\therefore$ $\{H^+\}_{EQ} = 5.62\cdot10^{-5}$ let X = ml 6.0M HCl added

Assume that the volume of HCl added (X) is negligible compared to the initial volume of the solution (100ml). Let Y = the molarity of HCl in the solution let $Z = -\Delta\{H^+\}$ (change step)

conc.	initial	added:	change	EQUILIBRIUM	$\therefore$
H^+	10^{-7}	Y	- Z	10^{-7}+Y-Z	$5.62\cdot10^{-5}$*
$C_2H_3O_2^-$	0.10		- Z	0.10 - Z	0.10 - Z
$HC_2H_3O_2$	0		+ Z	Z	Z

*This value is known since the equilibrium pH is specified.

$$\frac{\{H^+\}\{C_2H_3O_2^-\}}{\{HC_2H_3O_2\}} = 1.8\cdot10^{-5} = \frac{(5.62\cdot10^{-5})(0.10 - Z)}{Z}$$

$7.42\cdot10^{-5}Z = 5.62\cdot10^{-6}$ $\underline{Z = 7.57\cdot10^{-2}}$ Since: $10^{-7} + Y - Z = 5.62\cdot10^{-5}$ $\underline{Y} = 5.62\cdot10^{-5} + 7.57\cdot10^{-2} - 10^{-7} = \underline{7.58\cdot10^{-2}}$

$$\text{X = ?ml 6.0M HCl} = 100\text{ml soln}\cdot\frac{7.58\cdot10^{-2}\text{mol HCl}}{10^3\text{*ml soln}}\cdot\frac{10^3\text{*ml 6.0M HCl}}{6.0\text{mol HCl}} = \underline{\underline{1.3\text{ml}}}$$

(assumption OK)

15.79 $H_2CO_3 \rightleftharpoons H^+ + HCO_3^-$ $\frac{\{H^+\}\{HCO_3^-\}}{\{H_2CO_3\}} = 4.3\cdot10^{-7}$

pH = 7.43 $\underline{\{H^+\}_{EQ} = 3.72\cdot10^{-8}}$ $\{CO_2\} = 2.6\cdot10^{-2} = \{H_2CO_3\}$

$$\frac{(3.72\cdot10^{-8})\{HCO_3^-\}}{(2.6\cdot10^{-2})} = 4.3\cdot10^{-7}$$

$\underline{\underline{\{HCO_3^-\} = 0.30\text{M}}}$ (2 s.f.)

15.80 Since the K_a of $HNO_2 = 4.5 \cdot 10^{-4}$ while the K_b of $NH_3 = 1.8 \cdot 10^{-5}$ NH_4^+ is a stronger acid than NO_2^- is a base. Thus the behavior of 0.10M NH_4NO_2 can be described by the following equations:

(1) $NH_4^+ + NO_2^- \rightleftharpoons NH_3 + HNO_2$ (2) $NH_4^+ \rightleftharpoons H^+ + NH_3$

$$K_{(1)} = \frac{\{NH_3\}\{HNO_2\}}{\{NH_4^+\}\{NO_2^-\}} = \frac{\{NH_3\}}{\{NH_4^+\}\{OH^-\}} \bullet \frac{\{HNO_2\}\{OH^-\}}{\{NO_2^-\}} = \frac{K_{hy}(NO_2^-)}{K_b(NH_3)} = \frac{10^{-14}}{1.8 \bullet 10^{-5} \bullet 4.5 \bullet 10^{-4}}$$

$= 1.23 \bullet 10^{-6}$

$$K_{(2)} = K_{hy}(NH_4^+) = \frac{\{H^+\}\{NH_3\}}{\{NH_4^+\}} = 5.56 \bullet 10^{-10}$$

let $X = \Delta\{HNO_2\}$ (change 1) let $Y = \Delta\{H^+\}$ (change 2)

conc.	initial	chng 1	chng 2	EQUILIBRIUM	assume:	$\therefore$
NH_4^+	0.10	- X	- Y	0.10-X-Y	0.10>>X	0.10
NO_2^-	0.10	- X		0.10 - X	X>>Y	0.10
NH_3	0	+ X	+ Y	X + Y		X
HNO_2	0	+ X		X		X
H^+	10^{-7}		+ Y	10^{-7} + Y		$\{H^+\}$

$$K_{(1)} = 1.23 \bullet 10^{-6} = \frac{(X)(X)}{(0.10)(0.10)}$$ $X^2 = 1.23 \bullet 10^{-8}$ $X = 1.11 \bullet 10^{-4}$ (assumption 1 valid)

$$K_{(2)} = 5.56 \bullet 10^{-10} = \frac{(1.11 \bullet 10^{-4})\{H^+\}}{(0.10)}$$ $\{H^+\} = 5.01 \bullet 10^{-7} = 10^{-7} + Y$

$\therefore$ $Y = 4.01 \bullet 10^{-7}$ (assumption 2 valid) pH = 6.30

15.81

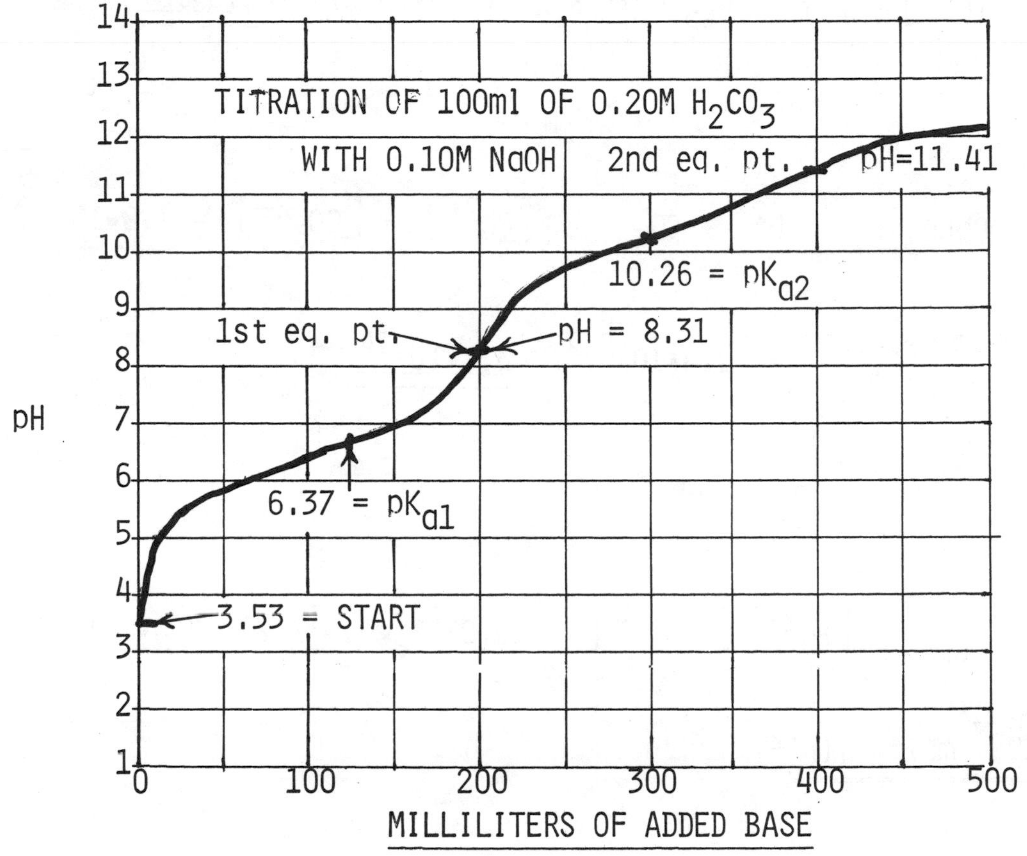

TITRATION OF 100ml OF 0.20M H_2CO_3
WITH 0.10M NaOH
2nd eq. pt.
pH=11.41
10.26 = pK_{a2}
1st eq. pt.
pH = 8.31
6.37 = pK_{a1}
3.53 = START
pH
14
13
12
11
10
9
8
7
6
5
4
3
2
1
0
100
200
300
400
500
MILLILITERS OF ADDED BASE

PROBLEM SOLUTIONS FOR CHAPTER 16 of BRADY & HUMISTON 3rd Ed.

16.11 $[Cu^{+}][Cl^{-}] = K_{sp} = ?$ $K_{sp} = (1.0 \cdot 10^{-3})^2 = \underline{\underline{1.0 \cdot 10^{-6}}}$

16.12 $PbCO_{3(s)} = Pb^{2+} + CO_3^{2-}$ $[Pb^{2+}] = [CO_3^{2-}] = \underline{1.8 \cdot 10^{-7}}$

$K_{sp} = [Pb^{2+}][CO_3^{2-}] = (1.8 \cdot 10^{-7})^2 = \underline{\underline{3.2 \cdot 10^{-14}}}$

16.13 $K_{sp} = [Ba^{2+}][C_2O_4^{2-}] = (0.0781 \text{ g} \cdot l^{-1} / 225.4 \text{ g} \cdot mol^{-1})^2 =$

$\underline{\underline{1.20 \cdot 10^{-7} \text{ mol}^2 \cdot ltr^{-2}}}$

16.14 $CaCrO_{4(s)} = Ca^{2+} + CrO_4^{2-}$ $K_{sp} = [Ca^{2+}][CrO_4^{2-}]$

$[Ca^{2+}] = [CrO_4^{2-}] = 1.0 \cdot 10^{-2}$ $\therefore$ $\underline{\underline{K_{sp}}} = (1.0 \cdot 10^{-2})^2 = \underline{\underline{1.0 \cdot 10^{-4}}}$

16.15 $K_{sp} = [Pb^{2+}][I^-]^2 = (1.5\cdot10^{-3})(2^* \times 1.5\cdot10^{-3})^2 = \underline{\underline{1.4\cdot10^{-8}}}$

16.16 $PbF_{2(s)} = Pb^{2+} + 2F^-$ $\quad K_{sp} = [Pb^{2+}][F^-]^2 = ?$

$$?M = ?\ \text{mol } PbF_2 = 1 \cancel{\text{l soltn}} \times \frac{0.0981\cancel{\text{g } PbF_2}}{0.200^* \cancel{\text{l soltn}}} \times \frac{1 \text{ mol } PbF_2}{245.2\cancel{\text{g } PbF_2}} = \underline{2.00\cdot10^{-3}M}$$

$$\therefore K_{sp} = (2.00\cdot10^{-3})(4.00\cdot10^{-3})^2 = \underline{\underline{3.20\cdot10^{-8}}} \quad ([F^-] = 2 \times [Pb^{2+}])$$

16.17 $?M = ?\ \text{mol } MgF_2 = 1^* \cancel{\text{l soltn}} \times \dfrac{7.6\cdot10^{-2}\text{g } \cancel{MgF_2}}{1^* \cancel{\text{l soltn}}} \times \dfrac{1^* \text{mol } MgF_2}{62.31\text{g } \cancel{MgF_2}} =$

$\underline{1.2\cdot10^{-3}M}$ $\quad K_{sp} = [Mg^{2+}][F^-]^2 = (1.2\cdot10^{-3})(2 \times 1.2\cdot10^{-3})^2 = \underline{\underline{6.9\cdot10^{-9}}}$

16.18 $?M\,Bi_2S_3 = ?\text{mol } Bi_2S_3 = 1^* \cancel{\text{ltr soln}} \cdot \dfrac{2.5\cdot10^{-12}\cancel{\text{g } Bi_2S_3}}{1^* \cancel{\text{ltr soln}}} \cdot \dfrac{1^* \text{mol } Bi_2S_3}{514.2\cancel{\text{g } Bi_2S_3}}$

$= \underline{4.86\cdot10^{-15}M\ Bi_2S_3}$ $\quad \underline{Bi_2S_3} \rightleftharpoons 2Bi^{3+} + 3S^{2-}$

(16.18 continued) $K_{sp} = \{Bi^{3+}\}^2\{S^{2-}\}^3$ Since 11mole of Bi_2S_3 yields 2 moles of Bi^{3+} and 3 moles of S^{2-}: $\{Bi^{3+}\}_{EQ} = 2 \bullet 4.86 \bullet 10^{-15} M$

$\{S^{2-}\}_{EQ} = 3 \bullet 4.86 \bullet 10^{-15} M$

$$K_{sp} = (9.72 \bullet 10^{-15})^2 (1.46 \bullet 10^{-14})^3 = \underline{\underline{2.9 \bullet 10^{-70}}}$$

16.19 $Ni(OH)_2 \rightleftharpoons Ni^{2+} + 2OH^-$ $K_{sp} = \{Ni^{2+}\}\{OH^-\}^2$

since: pH = 8.83 then: $\{OH^-\} = 6.8 \bullet 10^{-6}$ since: $\{OH^-\} = 2 \bullet \{Ni^{2+}\}$ then:

$K_{sp} = (6.8 \bullet 10^{-6}/2)(6.8 \bullet 10^{-6})^2 = \underline{\underline{1.6 \bullet 10^{-16}}}$

16.20 $MgC_2O_4 \rightleftharpoons Mg^{2+} + C_2O_4^{2-}$ $K_{sp} = \{Mg^{2+}\}\{C_2O_4^{2-}\}$

$$?M\ MgC_2O_4 = ?mol\ MgC_2O_4 = 1^* \cancel{1tr\ soln} \bullet \frac{0.47 \cancel{g\ MgC_2O_4}}{0.500 \cancel{1tr\ soln}} \bullet \frac{1^* mol\ MgC_2O_4}{112.3 \cancel{g\ MgC_2O_4}} =$$

$8.37 \bullet 10^{-3} M\ MgC_2O_4$

conc.	initial	added:	change	EQUILIBRIUM
Mg^{2+}	0	$8.37 \bullet 10^{-3}$		$8.37 \bullet 10^{-3}$
$C_2O_4^{2-}$	0.0020	$8.37 \bullet 10^{-3}$		$1.04 \bullet 10^{-2}$

$K_{sp} = (8.37 \bullet 10^{-3})(1.04 \bullet 10^{-12}) = \underline{\underline{8.7 \bullet 10^{-5}}}$

16.21 FOR EACH PART: let X = the molar solubility

(a) $PbS_{(s)} \rightleftharpoons Pb^{2+} + S^{2-}$ $K_{sp} = 7 \cdot 10^{-27} = \{Pb^{2+}\}\{S^{2-}\}$

$X^2 = 7 \cdot 10^{-27}$ $X = 8 \cdot 10^{-14} M$ PbS

(b) $Fe(OH)_{2(s)} \rightleftharpoons Fe^{2+} + 2OH^-$ $K_{sp} = \{Fe^{2+}\}\{OH^-\}^2 = 2 \cdot 10^{-15}$

conc.	initial	added:	change	EQUILIBRIUM	assume:	$\therefore$
OH^-	10^{-7}	2X		$10^{-7} + 2X$	$2X >> 10^{-7}$	2X
Fe^{2+}	0	X		X		X

$(X)(2X)^2 = 2 \cdot 10^{-15}$ $4X^3 = 2 \cdot 10^{-15}$ $X = 8 \cdot 10^{-6} M$

(c) $BaSO_{4(s)} \rightleftharpoons Ba^{2+} + SO_4^{2-}$ $K_{sp} = \{Ba^{2+}\}\{SO_4^{2-}\} = 1.5 \cdot 10^{-9}$

$X^2 = 1.5 \cdot 10^{-9}$ $X = 3.9 \cdot 10^{-5} M$ $BaSO_4$

(d) $Hg_2Cl_{2(s)} \rightleftharpoons Hg_2^{2+} + 2Cl^-$ $K_{sp} = \{Hg_2^{2+}\}\{Cl^-\}^2 = 2 \cdot 10^{-18}$

$(X)(2X)^2 = 2 \cdot 10^{-18}$ $4X^3 = 2 \cdot 10^{-18}$ $X = 8 \cdot 10^{-7} M$ Hg_2Cl_2

(e) $Al(OH)_{3(s)} \rightleftharpoons Al^{3+} + 3OH^-$ $K_{sp} = \{Al^{3+}\}\{OH^-\}^3 = 2 \cdot 10^{-33}$

conc.	initial	added:	change	EQUILIBRIUM	assume:	$\therefore$
OH^-	10^{-7}	3X		$3X + 10^{-7}$	$3X << 10^{-7}$	10^{-7}
Al^{3+}	0	X		X		X

$(X)(10^{-7})^3 = 2. \cdot 10^{-33}$ $X = 2 \cdot 10^{-12}$ (since $3X = 6 \cdot 10^{-12}$ the assumption was valid)

(16.21 continued) (f) $MgC_2O_{4(s)} \rightleftharpoons Mg^{2+} + C_2O_4^{2-}$

$K_{sp} = \{Mg^{2+}\}\{C_2O_4^{2-}\} = X^2 = 8.6\bullet10^{-5}$ $\qquad$ $\underline{\underline{X = 9.3\bullet10^{-3}M\ MgC_2O_4}}$

16.22 $Mg(OH)_{2(s)} \rightleftharpoons Mg^{2+} + 2OH^-$ $\qquad$ $K_{sp} = \{Mg^{2+}\}\{OH^-\}^2 = 1.2\bullet10^{-11}$

since: $\{OH^-\} = 2\bullet\{Mg^{2+}\}$ $\qquad$ $K_{sp} = \{OH^-\}/2 \bullet \{OH^-\}^2$ $\qquad$ $\underline{\{OH^-\} = 2.9\bullet10^{-4}}$

$\underline{\underline{pH = 10.46}}$

16.23 $CaSO_{4(s)} \rightleftharpoons Ca^{2+} + SO_4^{2-}$ $\qquad$ $K_{sp} = \{Ca^{2+}\}\{SO_4^{2-}\} = 2\bullet10^{-4}$

$\underline{\text{let } X = \text{molarity of } CaSO_4}$ $\qquad$ $X^2 = 2\bullet10^{-4}$ $\qquad$ $\underline{X = 1.4\bullet10^{-2}M\ CaSO_4}$

(only 1 s.f.)

$$?g\ CaSO_4 = 600\cancel{ml\ soln}\bullet\frac{1.4\bullet10^{-2}\cancel{mol\ CaSO_4}}{10^3{*}\cancel{ml\ soln}}\bullet\frac{136g\ CaSO_4}{1{*}\cancel{mol\ CaSO_4}} = \underline{\underline{1g\ CaSO_4}}$$

16.24 $HgS_{(s)} \rightleftharpoons Hg^{2+} + S^{2-}$ $\qquad$ $K_{sp} = \{Hg^{2+}\}\{S^{2-}\} = 1.6\bullet10^{-54}$

$\therefore$ $\underline{\{Hg^{2+}\} = 1.26\bullet10^{-27}M}$

$$?ltr\ soln = 1{*}\cancel{Hg^{2+}}\bullet\frac{1{*}1tr\ soln}{1.26\bullet10^{-27}\cancel{mol\ Hg^{2+}}}\bullet\frac{1{*}\cancel{mol\ Hg^{2+}}}{6.02\bullet10^{23}\cancel{Hg^{2+}}} = \underline{\underline{1300\ ltr\ soln}}$$

16.25 $CaCO_{3(s)} \rightleftharpoons Ca^{2+} + CO_3^{2-}$ $\quad K_{sp} = \{Ca^{2+}\}\{CO_3^{2-}\} = 9 \cdot 10^{-9}$

let X = M $CaCO_3$ $\therefore$ $\{Ca^{2+}\} = X$, $\{CO_3^{2-}\} = 0.50 + X$ $\quad$ assume: 0.50>>X

$(X)(0.50) = 9 \cdot 10^{-9}$ $\quad$ $X = 2 \cdot 10^{-8}$M $CaCO_3$

16.26 $AgCl_{(s)} \rightleftharpoons Ag^{+} + Cl^{-}$ $\quad K_{sp} = \{Ag^{+}\}\{Cl^{-}\} = 1.7 \cdot 10^{-10}$

let X = M AgCl $\quad$ 0.020M $AlCl_3 \longrightarrow$ 0.060M Cl^-

conc.	initial	added:	change	EQUILIBRIUM	assume:	$\therefore$
Cl^-	0.060	X		0.060 + X	0.06>>X	0.060
Ag^+	0	X		X		X

$(X)(0.060) = 1.7 \cdot 10^{-10}$ $\quad$ $X = 2.8 \cdot 10^{-9}$M AgCl

16.27 $PbCl_{2(s)} \rightleftharpoons Pb^{2+} + 2Cl^{-}$ $\quad K_{sp} = \{Pb^{2+}\}\{Cl^{-}\}^2 = 1.6 \cdot 10^{-5}$

0.020M $AlCl_3 \longrightarrow$ 0.060M Cl^- $\quad$ let X = M $PbCl_2$

conc.	initial	added:	change	EQUILIBRIUM	assume:	$\therefore$
Cl^-	0.060	2X		0.060 + 2X	0.06>>2X	0.060
Pb^{2+}	0	X		X		X

$(X)(0.060)^2 = 1.6 \cdot 10^{-5}$ $\quad$ $X = 4.4 \cdot 10^{-3}$M $PbCl_2$ (assumption shaky!)

(second approximation)

$(X)(0.060+0.0088)^2 = 1.6 \cdot 10^{-5}$ $\quad$ $X = 3.4 \cdot 10^{-3}$M $PbCl_2$ (a 3rd approximation gives $3.6 \cdot 10^{-3}$)

16.28 $Ag_2CrO_{4(s)} \rightleftharpoons 2Ag^+ + CrO_4^{2-}$ $K_{sp} = \{Ag^+\}^2\{CrO_4^{2-}\} = 1.9 \cdot 10^{-12}$

let X = M Ag_2CrO_4

conc.	initial	added:	Change	EQUILIBRIUM	ASSUME:	∴
Ag^+	0.10	2X		0.10 + 2X	0.1>>2X	0.10
CrO_4^{2-}	0	X		X		X

$(0.10)^2(X) = 1.9 \cdot 10^{-12}$ $X = 1.9 \cdot 10^{-10}$M $AgCrO_4$ (assumption valid)

16.29 $Ag_2CrO_{4(s)} \rightleftharpoons 2Ag^+ + CrO_4^{2-}$ $K_{sp} = \{Ag^+\}^2\{CrO_4^{2-}\} = 1.9 \cdot 10^{-12}$

let X = M Ag_2CrO_4

conc.	initial	added:	change	EQUILIBRIUM	Assume:	∴
Ag^+	0	2X		2X	0.1>>X	2X
CrO_4^{2-}	0.10	X		0.10 + X		0.10

$(2X)^2(0.10) = 1.9 \cdot 10^{-12}$ $4X^2 = 1.9 \cdot 10^{-11}$ $X = 2.2 \cdot 10^{-6}$M Ag_2CrO_4

(assumption valid)

16.30 $CaF_{2(s)} \rightleftharpoons Ca^{2+} + 2F^-$ $K_{sp} = \{Ca^{2+}\}\{F^-\}^2 = 1.7 \cdot 10^{-10}$

let X = M CaF_2

conc.	initial	added:	change	EQUILIBRIUM	assume:	∴
Ca^{2+}	0	X		X		X
F^-	0.010	2X		0.010 + 2X	0.01>>2X	0.010

(16.30 continued) $(X)(0.010)^2 = 1.7\bullet10^{-10}$ $X = 1.7\bullet10^{-6}M\ CaF_2$
(assumption valid)

16.31 $BaF_{2(s)} \rightleftharpoons Ba^{2+} + 2F^-$ $K_{sp} = \{Ba^{2+}\}\{F^-\}^2 = 1.7\bullet10^{-6}$

let X = M NaF

conc.	initial	added:	change	EQUILIBRIUM	assume:	∴
Ba^{2+}	0	$6.8\bullet10^{-4}$		$6.8\bullet10^{-4}$		$6.8\bullet10^{-4}$
F^-	X	$1.4\bullet10^{-3}$		$X + 1.4\bullet10^{-3}$	X>>.001	X

$(6.8\bullet10^{-4})(X)^2 = 1.7\bullet10^{-6}$ $X^2 = 2.5\bullet10^{-3}$ $X = 0.050M$ NaF
(assumption valid)

$?g\ NaF = 1*ltr\ soln\bullet\frac{0.050mol\ NaF}{1*ltr\ soln}\bullet\frac{42.0g\ NaF}{1*mol\ NaF} =$ 2.1g NaF per ltr soln

16.32 $AgC_2H_3O_{2(s)} \rightleftharpoons Ag^+ + C_2H_3O_2^-$ $K_{sp} = \{Ag^+\}\{C_2H_3O_2^-\} = 2.3\bullet10^{-3}$

$(5.0\bullet10^{-2})(1.0\bullet10^{-3}) = 5.0\bullet10^{-5}$ Since this ion product is smaller than the K_{sp} $(2.3\bullet10^{-3})$ no precipitate will form.

(b) $K_{sp} = \{Ba^{2+}\}\{F^-\}^2 = 1.7\bullet10^{-6}$ ion product $= (1.0\bullet10^{-2})(2.0\bullet10^{-2})^2 = 4.4\bullet10^{-6}$ Since this ion product is larger than the K_{sp} $(1.7\bullet10^{-6})$ some precipitate will form.

(c) $K_{sp} = \{Ca^{2+}\}\{SO_4^{2-}\} = 2\bullet10^{-4}$ ion product $= (500/750)(1.4\bullet10^{-2})\bullet(250/750)(0.25) = 7.8\bullet10^{-4}$ Since ion product $>K_{sp}$ precipitate forms.

16.33 $Fe(OH)_{2(s)} \rightleftharpoons Fe^{2+} + 2OH^-$ $K_{sp}=\{Fe^{2+}\}\{OH^-\}^2 = 2\cdot10^{-15}$

$(0.010)\{OH^-\}^2 = 2\cdot10^{-15}$ $\{OH^-\}^2 = 2\cdot10^{-13}$ $\{OH^-\} = 4.5\cdot10^{-7}$ pH=7.65

16.34 $AgCl_{(s)} \rightleftharpoons Ag^+ + Cl^-$ $K_{sp}=\{Ag^+\}\{Cl^-\} = 1.7\cdot10^{-10}$

let $X = -\Delta\{Ag^+\}$ (change step)

conc.	initial	after mix	change	EQUILIBRIUM	assume:	∴
Ag^+	0.20	0.10	-X	0.10 - X	X≈0.050	0.050
Cl^-	0.10	0.050	-X	0.050 - X		$\{Cl^-\}$

$(0.050)\{Cl^-\} = 1.7\cdot10^{-10}$ $\{Cl^-\} = 3.4\cdot10^{-9}M$ (assumption valid)

$\{Ag^+\} = 0.050M$ $\{NO_3^-\} = 0.10M$ $\{H^+\} = 0.050$

16.35 (a) $K_{sp}=\{Ca^{2+}\}\{CO_3^{2-}\} = 9\cdot10^{-9}$ $(0.025)(0.0050) = 1.3\cdot10^{-4}$

Since the ion product is larger than the K_{sp} precipitate will form.

(b) $K_{sp}=\{Pb^{2+}\}\{Cl^-\}^2 = 1.6\cdot10^{-5}$ ion product = $(0.010)(0.060)^2$ = $3.6\cdot10^{-5}$ larger than K_{sp} so precipitate forms.

(c) $K_{sp}=\{Fe^{2+}\}\{C_2O_4^{2-}\} = 2.1\cdot10^{-7}$ ion product = $(1.5\cdot10^{-3})(2.2\cdot10^{-3})$= $3.3\cdot10^{-6}$ larger than K_{sp} so precipitate forms.

16.36 $K_{sp}(PbCrO_4) = 1.8 \cdot 10^{-14}$ $K_{sp}(BaCrO_4) = 2.4 \cdot 10^{-10}$ The CrO_4^{2-} must exceed $1.8 \cdot 10^{-12}$ to ppt $PbCrO_4$. (only $PbCrO_4$ ppt untill the CrO_4^{2-} concentration reaches $2.4 \cdot 10^{-8}$)(At this point the Pb^{2+} concentration is reduced to $7.5 \cdot 10^{-7}$) and precipitation of $BaCrO_4$ begins.)

16.37 $K_{sp}(PbS) = 7 \cdot 10^{-27}$ $K_{sp}(NiS) = 2 \cdot 10^{-21}$ For the more soluable NiS the maximum S^{2-} concentration allowable without precipitation $= 2 \cdot 10^{-21}/0.010 = \underline{2 \cdot 10^{-19}}$

$$1.1 \cdot 10^{-21} = \frac{\{H^+\}^2\{S^{2-}\}}{H_2S} = \frac{\{H^+\}^2(2 \cdot 10^{-19})}{(0.10)}$$

$\underline{\{H^+\} = 0.023}$ $\underline{\underline{pH = 1.64}}$ (Actually the pH should be kept slightly lower to assure that no NiS precipitates.)

16.38 $K_{sp}(FeS) = 3.7 \cdot 10^{-19}$ $K_{sp}(ZnS) = 1.2 \cdot 10^{-23}$

The maximum$\{S^{2-}\}$ without precipitation of FeS $= 3.7 \cdot 10^{-19}/0.10 = \underline{3.7 \cdot 10^{-18}}$

$$\frac{\{H^+\}^2\{S^{2-}\}}{\{H_2S\}} = 1.1 \cdot 10^{-21} = \frac{\{H^+\}^2(3.7 \cdot 10^{-18})}{(0.10)}$$

$\underline{\underline{\{H^+\} = 5.5 \cdot 10^{-3}}}$ (or slightly higher)

$\{Zn^{2+}\}(3.7 \cdot 10^{-18}) = 1.2 \cdot 10^{-23}$ $\underline{\underline{\{Zn^{2+}\} = 3.2 \cdot 10^{-6}M}}$

16.39 $K_{sp}=\{Hg^{2+}\}\{S^{2-}\} = 1/6 \cdot 10^{-54}$ $\{S^{2-}\}_{max} = 1.6 \cdot 10^{-54}/0.0010 =$ $\underline{1.6 \cdot 10^{-51}}$ To reduce $\{S^{2-}\}$ to this would require ******WOW!******** $\underline{\underline{H^+ = 26000000000000M}}$

16.40 $K_{sp}=\{Zn^{2+}\}\{S^{2-}\} = 1.2\bullet10^{-23}$ $\dfrac{\{H^+\}^2\{S^{2-}\}}{\{H_2S\}} = 1.1\bullet10^{-21}$

if $\{H^+\} = 12M$ then: $\dfrac{(12)^2\ S^{2-}}{(0.10)} = 1.1\bullet10^{-21}$ $\underline{\{S^{2-}\} = 7.6\bullet10^{-25}}$

$\{Zn^{2+}\}(7.6\bullet10^{-25}) = 1.2\bullet10^{-23}$ $\underline{\{Zn^{2-}\} = 16M}$ $\therefore$ To precipitate ZnS, the $\{Zn^{2+}\}$ would have to exceed 16M which is a practical impossibility, so ZnS is soluble in concentrated HCl.

16.41 $8.6\bullet10^{-5} = \{Mg^{2+}\}\{C_2O_4^{2-}\}$ $2.3\bullet10^{-9} = \{Ca^{2+}\}\{C_2O_4^{2-}\}$

$\dfrac{\{H^+\}^2\{C_2O_4^{2-}\}}{\{H_2C_2O_4\}} = 6.5\bullet10^{-2}\bullet6.1\bullet10^{-5} = 4.0\bullet10^{-6} = \dfrac{\{H^+\}^2(8.6\bullet10^{-5}/0.10)}{(0.10)}$

$\{H^+\}^2 = 4.65\bullet10^{-4}$ $\underline{\{H^+\} = 0.022M}$ pH = 1.66 At this pH the solution will be saturated with respect to MgC_2O_4 and the Ca^{2+} concentration will be reduced to $2.7\bullet10^{-6}M$. This represents maximum separation since 99.997% of the original Ca^{2+} has been precipitated while all of the Mg^{2+} remains in solution.

16.42 $K_{sp} = Ag^+\ I^- = 8.5\bullet10^{-17}$ $K_{inst} = \dfrac{\{Ag^+\}\{CN^-\}^2}{\{Ag(CN)_2^-\}} = 1.9\bullet10^{-19}$

let X = moles of AgI which would dissolve per liter of solution

let Y = $\Delta\{Ag(CN)_2^-\}$ (during the change step)

(16.42 continued)

conc.	initial	added:	change	EQUILIBRIUM	assume:	∴
Ag^+	0	X	- Y	X - Y	X≈Y	Ag^+
CN^-	0.010	0	- 2Y	0.010 - 2Y	2Y≈0.01	CN^-
$Ag(CN)_2^-$	0	0	+ Y	Y	Y≈0.005	0.005
I^-	0	X	0	X		0.005

$8.5\cdot10^{-17} = \{Ag^+\}(0.005)$ $\{Ag^+\} = 1.7\cdot10^{-14}M$ $\dfrac{(1.7\cdot10^{-14})\{CN^-\}^2}{(0.005)} = 1.9\cdot10^{-19}$

$\{CN^-\} = 2.4\cdot10^{-4}M$ $\{I^-\} = 0.0050M = \{Ag(CN)_2^-\}$

since: $\{CN^-\}_{EQ} = 0.010 - 2Y$ $Y = (0.010 - 2.4\cdot10^{-4})/2 = 4.9\cdot10^{-3} = X$

molar solubility of AgI in 0.010M KCN = $4.9\cdot10^{-3}$

16.43 $Zn(OH)_{2(s)} \rightleftharpoons Zn^{2+} + 2OH^-$ $K_{sp} = \{Zn^{2+}\}\{OH^-\}^2 = 4.5\cdot10^{-17}$

$Zn(NH_3)_4^{2+} \rightleftharpoons Zn^{2+} + 4NH_3$ $\dfrac{\{Zn^{2+}\}\{NH_3\}^4}{\{Zn(NH_3)_4^{2+}\}} = K_{inst}$

let $X = \Delta\{Zn(NH_3)_4^{2+}\}$ (change step)

conc.	initial	added:	change	EQUILIBRIUM	assume:	∴
Zn^{2+}	$5.7\cdot10^{-3}$	0	- X	$5.7\cdot10^{-3}$-X	$X\approx5.7\cdot10^{-3}$	Zn^{2+}
OH^-	0.011	0	0	0.011		0.011
NH_3	1.0	0	- 4X	1.0 - 4X	1.0>>4X	1.0
$Zn(NH_3)_4^{2+}$	0	0	+ X	X		0.0057

$[Zn^{2+}](0.011)^2 = 4.5\cdot10^{-17}$ $[Zn^{2+}] = 3.7\cdot10^{-13}M$ (1st assumption valid)

(16.43 continued) $\dfrac{(3.7\cdot10^{-13})(1.0)}{(0.0057)} = K_{inst} = 6.5\cdot10^{-11}$

16.44 $K_{sp} = \{Mg^{2+}\}\{OH^-\}^2 = 1.2\cdot10^{-11}$ $\qquad K_b = 1.8\cdot10^{-5} = \dfrac{\{NH_4^+\}\{OH^-\}}{\{NH_3\}}$

let X = molar solubility of $Mg(OH)_2$ $\qquad$ let Y = $\Delta\{OH^-\}$ (change step)

conc.	initial	added:	change	EQUILIBRIUM	assume:	∴
Mg^{2+}	0	X		X	$Y \gg 10^{-7}$	X
OH^-	10^{-7}	2X	+ Y	$10^{-7}+2X+Y$	$Y \gg 2X$	Y
NH_4^+			+ Y	Y		Y
NH_3	0.10		- Y	0.10 - Y	$0.10 \gg Y$	0.10

$\dfrac{(Y)(Y)}{(0.10)} = 1.8\cdot10^{-5}$ $\qquad$ $Y = 1.3\cdot10^{-3}$ (1st and 3rd assumptions are valid)

$(X)(1.3\cdot10^{-3})^2 = 1.2\cdot10^{-11}$ $\qquad$ $X = 7.1\cdot10^{-6}$ (2nd assumption OK)

molar solubility of $Mg(OH)_2 = 7.1\cdot10^{-6}$

16.45 $HC_2H_3O_2 \rightleftharpoons H^+ + C_2H_3O_2^-$ $\qquad K_a = 1.8\cdot10^{-5} = \dfrac{\{H^+\}\{C_2H_3O_2^-\}}{\{HC_2H_3O_2\}}$

$K_{sp} = 2.3\cdot10^{-3} = \{Ag^+\}\{C_2H_3O_2^-\}$ $\qquad$ let X = $\Delta\{C_2H_3O_2^-\}$ (change step)

In this problem we will not deal with equilibrium concentrations since we are concerned with the state of the system before precipitatio

(16.45 continued)

conc.	initial	added:	change	(before pptn)	assume:	∴
$HC_2H_3O_2$	1.0		- X	1.0 - X	1.0>>X	1.0
H^+	10^{-7}		+ X	10^{-7} + X	X>>10^{-7}	X
$C_2H_3O_2^-$			+ X	X		X
Ag^+	1.0			1.0		1.0

$\frac{(X)(X)}{(1.0)} = 1.8\cdot10^{-5}$ $\underline{X = 4.2\cdot10^{-3}}$ (assumptions valid)

ION PRODUCT = $(1.0)(4.2\cdot10^{-3}) = 4.2\cdot10^{-3}$ Since the ion product is larger than the K_{sp} ($2.3\cdot10^{-3}$) precipitation will occur.

16.46 (See 16.45 for reference data.) let X = $mol\cdot ltr^{-1}$ $NaC_2H_3O_2$

conc.	initial	added:	change	EQUILIBRIUM	assume:	∴
Ag^+	0.200	0		0.200		0.200
$C_2H_3O_2^-$	0	X	- Y	X - Y	X>Y	$\{C_2H_3O_2^-\}$
H^+	0.100		- Y	0.100 - Y	0.1>Y	$\{H^+\}$
$HC_2H_3O_2$			+ Y	Y		Y

NOTE: Y = $\Delta\{HC_2H_3O_2\}$ (change step) $(0.200)\{C_2H_3O_2^-\} = 2.3\cdot10^{-3}$

$\{C_2H_3O_2^-\}$ = 0.0115M (SATURATION VALUE) (Since this value equals X - Y the first assumption is valid.)

$\frac{(0.100-Y)(0.0115)}{(Y)} = 1.8\cdot10^{-5}$ Y = 0.0998 (2nd assumption valid)

since 0.0115 = X - Y then $\underline{X = 0.0115 + 0.0998 = 0.111}$

(16.46 continued)

$$?g\ NaC_2H_3O_2 = 0.200\ \cancel{ltr\ soln} \cdot \frac{0.111\,mol\ \cancel{NaC_2H_3O_2}}{1^*\cancel{ltr\ soln}} \cdot \frac{82.04g\ NaC_2H_3O_2}{1^*\cancel{mol\ NaC_2H_3O_2}} = \underline{1.8g}$$

In order to cause precipitation the amount of $NaC_2H_3O_2$ added would have to EXCEED 1.8 grams.

16.47 (See 16.45 for reference data.) $HF \rightleftharpoons H^+ + F^-$

$$K_a = 6.5 \cdot 10^{-4} = \frac{\{H^+\}\{F^-\}}{\{HF\}}$$

let X = $mol \cdot ltr^{-1}$ KF added

let Y = $\Delta\{H^+\}$ (due to $HC_2H_3O_2$ ionization)

let Z = $\Delta\{HF\}$ (change step)

conc.	initial	added:	change	EQUILIBRIUM	assume:	∴
Ag^+	0.20			0.20		0.20
$HC_2H_3O_2$	0.10		- Y	0.10 - Y		0.10-Y
$C_2H_3O_2^-$			+ Y	Y		Y
F^-		X	- Z	X - Z		X - Z
H^+	10^{-7}		+Y -Z	10^{-7}+Y -Z	$Z \gg 10^{-7}$	Y - Z
HF			+ Z	Z		Z

$(0.20)(Y) = 2.3 \cdot 10^{-3}$ $\quad$ $Y = 0.0115M = \{C_2H_3O_2^-\}$ (saturation value)

$$\frac{(Y - Z)(Y)}{(0.10-Y)} = 1.8 \cdot 10^{-5} = \frac{(0.0115-Z)(0.0115)}{(0.10-0.0115)}$$

(the assumption was OK)

$Z = 0.0114 = \{HF\}$

$$\frac{(Y - Z)(X - Z)}{(Z)} = 6.5 \cdot 10^{-4}$$

Although we have determined both Y and Z, the value of (Y-Z) = $\{H^+\}$ can not be determined precisely by taking the difference, since this would yield a quantity with only one s.f. By utilizing the acetic acid equilibrium again and solving for (Y-Z) as an entity, we obtain $1.39 \cdot 10^{-4}$ (precise to 2 s.f.).

(16.47 continued) Using the HF equilibrium as expressed on the last page and substituting the value of Z and the value of (Y-Z) just obtained:

$$\frac{(1.39\cdot10^{-4})(X - 0.0114)}{(0.0114)} = 6.5\cdot10^{-4}$$

$X = 6.47\cdot10^{-2}$ = $mol\cdot ltr^{-1}$ KF added

$$?g\ NaF = 0.200\,\text{ltr soln}\cdot\frac{0.0647\,\text{mol KF}}{1^*\,\text{ltr soln}}\cdot\frac{58.1\,\text{g KF}}{1^*\,\text{mol KF}} = 0.75\,\text{g KF}$$

(To cause precipitation of $AgC_2H_3O_2$ the amount of KF added must exceed this value.)

16.48 $K_{sp} = 3.7\cdot10^{-19} = \{Fe^{2+}\}\{S^{2-}\}$ Since: $\{Fe^{2+}\}_{EQ} = 0.20M$ ∴

$\{S^{2-}\}_{EQ} = 1.85\cdot10^{-18}$

$$\frac{\{H^+\}^2\{S^{2-}\}}{\{H_2S\}} = 1.1\cdot10^{-21} = \frac{\{H^+\}^2(1.85\cdot10^{-18})}{(0.10)}$$

$\{H^+\} = 7.7\cdot10^{-3}$ An additional $2\cdot0.20 mol\cdot ltr^{-1}$ of H^+ was used up by reaction with S^{2-} to form H_2S (some of which would escape from solution as a gas) so the total moles per liter of HCl which must be added to dissolve the 0.20 moles per liter of FeS = 0.41

16.49 $K_{sp} = 1.2\cdot10^{-11} = \{Mg^{2+}\}\{OH^-\}^2$

$$\frac{\{NH_3\}}{\{NH_4^+\}\{OH^-\}} = \frac{1}{K_b} = \frac{1}{1.8\cdot10^{-5}}$$

let X = $mol\cdot ltr^{-1}$ NH_4Cl added let Y = $\Delta\{NH_3\}$

(In order to avoid confusion, the completed table will be shown on the following page instead of portions on both pages.)

(16.49 continued)

conc.	initial	added:	change	EQUILIBRIUM	assume:	$\therefore$
NH_4^+		X	- Y	X - Y	Y≈0.20 (almost all of the OH^- must be consumed)	X-0.20
Mg^{2+}	0.10			0.10		0.10
OH^-	0.20		- Y	0.20 - Y		OH^-
NH_3			+ Y	Y		0.20

$(0.10)\{OH^-\}^2 = 1.2\cdot10^{-11}$ $\quad \underline{\{OH^-\} = 1.1\cdot10^{-5}}$ (assumption was valid)

$$\frac{(0.20)}{(X - 0.20)(1.1\cdot10^{-5})} = 5.56\cdot10^4 \qquad \underline{\underline{X = 0.53 = \text{mol}\cdot\text{ltr}^{-1}\ NH_4Cl \text{ added}}}$$

16.50 $K_{sp} = \{Ca^{2+}\}\{SO_4^{2-}\} = 2\cdot10^{-4}$ $\quad \underline{\{Ca^{2+}\} = \{SO_4^{2-}\} = 1.4\cdot10^{-2} = 0.014M\ CaSO_4}$

$$?\text{mol } CaSO_4 = 1^*\cancel{\text{hole}}\cdot\frac{(3.142\cdot0.5^2\cancel{cm^2}\cdot1.50\cancel{cm})}{1^*\cancel{\text{hole}}}\cdot\frac{0.97\cancel{g\ CaSO_4}}{1^*\cancel{cm^3\ CaSO_4}}\cdot\frac{1^*\text{mol } CaSO_4}{136\cancel{g\ CaSO_4}} = \underline{8.4\cdot10^{-3}\text{mol } CaSO_4} \text{ (in the hole)}$$

(only 1 s.f. - see K_{sp}

$$?\text{ days} = 8.4\cdot10^{-3}\cancel{\text{mol } CaSO_4}\cdot\frac{1^*\cancel{\text{ltr}}}{0.014\cancel{\text{mol } CaSO_4}}\cdot\frac{1^*\text{day}}{2.0\cancel{\text{ltr}}} = \underline{\underline{0.3 \text{ days}}}$$

16.51 $?\text{mol}Mg(OH)_2 = 0.025\cancel{\text{Ltr}}\cdot\frac{0.10\cancel{\text{mol HCl}}}{1^*\cancel{\text{ltr}}}\cdot\frac{1^*\text{mol } Mg(OH)_2}{2^*\cancel{\text{mol HCl}}} = \underline{1.25\cdot10^{-3}\text{mol}}$

$\therefore \underline{\underline{[Mg^{2+}] = 1.25\cdot10^{-3}\text{mol}/1.025\text{ltr} = 1.2\cdot10^{-3}M}}$

(16.51 continued) $Mg(OH)_{2(s)} \rightleftharpoons Mg^{2+} + 2OH^-$ $\{Mg^{2+}\}\{OH^-\}^2 = K_{sp}$

$= 1.2 \bullet 10^{-11} = (1.2 \bullet 10^{-3})\{OH^-\}^2$ $\{OH^-\} = 1.0 \bullet 10^{-4}$ pH = 10.00

16.52 $?M(OH^-) = ?mol\ OH^- = 1^*\ ltr\ soln \bullet \frac{2.20g\ NaOH}{0.250ltr\ soln} \bullet \frac{1^*mol\ NaOH}{40.0g\ NaOH} =$ $0.220M(OH^-)$

let $X = -\Delta\{Fe^{2+}\}$ (change step)

conc.	initial	added:	change	EQUILIBRIUM	assume:	∴
Fe^{2+}	0.10		- X	0.10 - X	X≈0.10*	Fe^{2+}
OH^-	0.220	0.220	- 2X	0.220 - 2X		≈0.020

*We have assumed that almost all of the Fe^{2+} present in the original solution is precipitated as $Fe(OH)_2$.

$K_{sp} = \{Fe^{2+}\}\{OH^-\}^2 = 2 \bullet 10^{-15} = \{Fe^{2+}\}(0.020)^2$ $\{Fe^{2+}\} = 5 \bullet 10^{-12}M$

$?g\ Fe(OH)_2 = 0.250ltr\ soln \bullet \frac{0.10mol\ Fe^{2+}}{1^*ltr\ soln} \bullet \frac{89.9g\ Fe(OH)_2}{1^*mol\ Fe^{2+}} = 2.2g\ Fe(OH)_2 ppt$

16.53 $Ni(OH)_{2(s)} \rightleftharpoons Ni^{2+} + 2OH^-$ $K_{sp} = \{Ni^{2+}\}\{OH^-\}^2 = 1.6 \bullet 10^{-14}$

$?M(OH^-) = ?mol\ OH^- = 1^*\ ltr\ soln \bullet \frac{1.75g\ NaOH}{0.250ltr\ soln} \bullet \frac{1^*mol\ OH^-}{40.0g NaOH} = 0.175M(OH^-)$

let $X = -\Delta\{Ni^{2+}\}$ (change step) (see table on the next page.)

(16.53 continued)

conc.	initial	added:	change	EQUILIBRIUM	assume:	∴
Ni^{2+}	0.10		- X	0.10 - X	2X≃0.175	0.0125
OH^-	10^{-7}	0.175	- 2X	0.175 - 2X		OH^-

We have assumed that almost all of the added OH^- was used up in the formation of $Ni(OH)_2$ since inspection of the initial concentration of Ni^{2+} indicates that OH^- is the limiting reactant. The alternative to this assumption was the solution of a cubic equation. (The solution of the cubic equation yields the same value for X, to 2 s.f., as we obtain below by the use of our approximation.)

$(0.0125)\{OH^-\}^2 = 1.6\cdot10^{-14}$ $\{OH^-\} = 1.13\cdot10^{-6}$ ∴ pH = 8.05

The moles of $Ni(OH)_2$ precipitated per liter of solution equal $-\Delta\{Ni^{2+}\} =$ X = 0.0875.

$$?g\ Ni(OH)_2 ppt = 0.250 \cancel{ltr\ soln}\cdot\frac{0.0875 \cancel{mol\ Ni(OH)_2} ppt}{1^* \cancel{ltr\ soln}}\cdot\frac{92.7 g\ Ni(OH)_2}{1^* \cancel{mol\ Ni(OH)_2}} =$$

2.0 g $Ni(OH)_2$ppt

16.54 $K_{sp}=\{Mn^{2+}\}\{OH^-\}^2 = 4.5\cdot10^{-14}$ $K_{sp}=\{Fe^{2+}\}\{OH^-\}^2 = 2\cdot10^{-15}$

let X = $mol\cdot ltr^{-1}$ $Mn(OH)_2$ added let Y = $-\Delta\{Fe^{2+}\}$ (change step)

conc.	initial	added:	change	EQUILIBRIUM	assume:	∴
Fe^{2+}	0.100		- Y	0.100 - Y	Y≃10^{-1}	Fe^{2+}
Mn^{2+}		X		X	X≃10^{-1}	0.100
OH^-	10^{-7}	2X	- 2Y	10^{-7}+2X-2Y		OH^-

$(0.100)\{OH^-\}^2 = 4.5\cdot10^{-14}$ $\{OH^-\} = 6.7\cdot10^{-7}$ $\{Fe^{2+}\}(6.7\cdot10^{-7})^2=2\cdot10^{-1}$

$\{Fe^{2+}\} = 4\cdot10^{-3}M$ pH = 7.83 $\{Mn^{2+}\} = 0.10M$

PROBLEM SOLUTIONS FOR CHAPTER 17 of BRADY & HUMISTON 3rd Ed.

17.38 (a) 2 (b) 1 (c) 5 (d) 2 (e) 8

17.39 $? \text{ e} = 1 \text{ C} \times \frac{1 \text{ F}}{96{,}500 \text{ C}} \times \frac{6.022 \cdot 10^{23} \text{ e}}{1 \text{ F}} = \underline{\underline{6.24 \cdot 10^{18} \text{e}}}$

17.40 (a) 1 (b) 4 (c) 10 (d) 2 (e) 8

17.41 (a) $? \text{ F} = 8950 \text{ C} \times \frac{1 \text{ F}}{96{,}500 \text{ C}} = \underline{\underline{0.0927 \text{ F}}}$

(b) $? \text{ F} = 1.5 \text{ A} \cdot 30 \text{ s} \times \frac{1 \text{ C} \cdot \text{s}}{1 \text{ A}} \times \frac{1 \text{ F}}{96{,}500 \text{ C}} = \underline{\underline{4.7 \cdot 10^{-4} \text{F}}}$ (c) $\underline{\underline{9.1 \cdot 10^{-2} \text{F}}}$

17.42 (a) $? \text{ min} = 10{,}500 \text{ C} \times \frac{1 \text{ s}}{25 \text{ C}} \times \frac{1 \text{ min}}{60 \text{ s}} = \underline{\underline{7.0 \text{ min}}}$

(17.42 continued)

(b) $? \text{ min} = 0.65\,\text{F} \times \frac{96{,}500\ \text{C}}{1\ \text{F}} \times \frac{1\ \text{s}}{15\ \text{C}} \times \frac{1\ \text{min}}{60\ \text{s}} = \underline{\underline{70\ \text{min}}}$ (2 s.f.)

(c) $? \text{ min} = 0.20\ \text{mol Cu}^{2+}/\text{Cu} \times \frac{2\ \text{F}}{1\ \text{mol Cu}^{2+}/\text{Cu}} \times \frac{96{,}500\ \text{C}}{1\ \text{F}} \times \frac{1\ \text{min}}{12 \times 60\ \text{C}} =$ $\underline{\underline{54\ \text{min}}}$

17.43 (a) $? \text{ min} = \frac{84{,}200\ \text{C}}{6.30\ \text{A}} \times \frac{1\ \text{A}}{1\ \text{C}\cdot\text{s}^{-1}} \times \frac{1\ \text{min}}{60\ \text{s}} = \underline{\underline{223\ \text{min}}}$

(b) $\underline{\underline{239\ \text{min}}}$ (c) $? \text{ min} = \frac{0.50\ \text{mol Al}}{18.3\ \text{A}} \times \frac{3 \times 96{,}500\ \text{C}}{1\ \text{mol Al}^{3+}} \times \frac{1\ \text{A}}{1\ \text{C}\cdot\text{s}^{-1}} \times \frac{1\ \text{min}}{60\ \text{s}}$ $= \underline{\underline{132\ \text{min}}}$

17.44 (a) $? \text{ F} = 10.0\ \text{ml O}_2\text{(STP)} \times \frac{1\ \text{mol O}_2}{22{,}400\ \text{ml O}_2\text{(STP)}} \times \frac{4\ \text{F}}{1\ \text{mol O}_2\text{(STP)}}$ $= \underline{\underline{1.79 \cdot 10^{-3}\text{F}}}$

(b) $?\text{F} = 10.0\ \text{g Al} \cdot \frac{1\ \text{mol Al}}{26.98\ \text{g Al}} \cdot \frac{3\ \text{F}}{1\ \text{mol Al}} = \underline{\underline{1.11\text{F}}}$

(17.44 continued) (c) $?F = 5.00\,\text{g Na} \cdot \dfrac{1^*\text{mol Na}}{22.99\,\text{g Na}} \cdot \dfrac{1^*F}{1^*\text{mol Na}} = \underline{\underline{0.217\ F}}$

(d) $?F = 5.00\,\text{g Mg} \cdot \dfrac{1^*\text{mol Mg}}{24.31\,\text{g Mg}} \cdot \dfrac{2^*F}{1^*\text{mol Mg}} = \underline{\underline{0.411\ F}}$

17.45 $?\text{g Na} = 25\,\text{A} \cdot 8.0\,\text{hr} \cdot \dfrac{3600^*\text{s}}{1^*\text{hr}} \cdot \dfrac{1^*\text{C}\cdot\text{s}^{-1}}{1^*\text{A}} \cdot \dfrac{23.0\,\text{g Na}}{96{,}500\,\text{C}} = \underline{\underline{170\text{g Na}}}$ $\underline{\underline{(260\text{g } Cl_2)}}$ (only 2 s.f.)

17.46 $?\text{g } O_2 = 0.50\,\text{A} \cdot 1.0\,\text{hr} \cdot \dfrac{1^*\text{C}\cdot\text{s}^{-1}}{1^*\text{A}} \cdot \dfrac{3600^*\text{s}}{1^*\text{hr}} \cdot \dfrac{8.00\,\text{g } O_2}{96{,}500\,\text{C}} = \underline{\underline{0.15\text{g } O_2}}$

$\underline{\underline{(0.019\text{g } H_2)}}$ $?\text{ltr } O_2(\text{STP}) = 0.15\,\text{g } O_2 \cdot \dfrac{22.4\,\text{ltr } O_2(\text{STP})}{32.00\,\text{g } O_2} = \underline{\underline{0.11\ \text{ltr } O_2(\text{STP})}}$

$\underline{\underline{0.21\ \text{ltr } H_2(\text{STP})}}$

17.47 $?\text{g Cu} = 115\,\text{A} \cdot 8.00\,\text{hr} \cdot \dfrac{1^*\text{C}\cdot\text{s}^{-1}}{1^*\text{A}} \cdot \dfrac{3600^*\text{s}}{1^*\text{hr}} \cdot \dfrac{63.54/2\,\text{gCu}}{96{,}500\,\text{C}} = \underline{\underline{1090\text{g Cu}}}$

17.48 $?\text{g Ag} = 8.00\,\text{hr} \cdot 8.46\,\text{A} \cdot \dfrac{1^*\text{C}\cdot\text{s}^{-1}}{1^*\text{A}} \cdot \dfrac{3600^*\text{s}}{1^*\text{hr}} \cdot \dfrac{107.9\,\text{g Ag}}{96{,}500\,\text{C}} = \underline{\underline{272\text{g Ag}}}$

$?\text{cm}^2\text{Ag} = \dfrac{272\,\text{g Ag}}{0.00254\,\text{cm Ag}} \cdot \dfrac{1^*\text{cm}^3\text{Ag}}{10.5\,\text{g Ag}} = \underline{\underline{1.02 \cdot 10^4\,\text{cm}^2\text{Ag}}}$

17.49 $?s = \frac{21.4 \cdot g\ Ag}{10.0\ A} \cdot \frac{1^*A}{1^*C \cdot s^{-1}} \cdot \frac{96,500\ C}{107.9 g\ Ag} = \underline{\underline{1910\ s}}$

17.50 $?hr = \frac{35.3 \cdot g\ Cr}{6.00\ A} \cdot \frac{1^*A}{1^*C \cdot s^{-1}} \cdot \frac{3^*F}{1^*mol\ Cr^{3+}} \cdot \frac{96,500\ C}{1^*F} \cdot \frac{1^*mol\ Cr}{52.00 g\ Cr} \cdot \frac{1^*hr}{3600 s}$
$= \underline{\underline{9.10 hr}}$

17.51 $?min = \frac{5.00 \cdot g\ Cu}{5.00\ A} \cdot \frac{1^*A}{1^*C \cdot s^{-1}} \cdot \frac{96,500\ C}{63.54/2\ gCu} \cdot \frac{1^*min}{60^*s} = \underline{\underline{50.6\ min}}$

17.52 $?A = \frac{0.225 \cdot g\ Ni}{10.0\ min} \cdot \frac{96,500\ C}{58.71/2\ gNi} \cdot \frac{1^*min}{60^*s} \cdot \frac{1^*A}{1^*C \cdot s^{-1}} = \underline{\underline{1.23\ A}}$

17.53 $?A = \frac{1.33 \cdot g\ Cl_2}{45.0\ min} \cdot \frac{96,500\ C}{35.45 g\ Cl_2} \cdot \frac{1^*min}{60^*s} \cdot \frac{1^*A}{1^*C \cdot s^{-1}} = \underline{\underline{1.34\ A}}$

17.54 (a) $?F = 1.25 \cdot g\ Cu \cdot \frac{1^*F}{63.54/2\ gCu} = \underline{\underline{0.0393\ F}}$

(b) $?gX = 1^*mol\ X \cdot \frac{3.42 g\ X}{0.0393 F} \cdot \frac{2^*F}{1^*mol\ X} = \underline{\underline{174\ g\ X\ \text{(atomic weight of X)}}}$

17.55 $?\text{mol Cr} = 0.125\cancel{\text{mol Cu}} \cdot \frac{2\cancel{F}}{1\cancel{\text{mol Cu}}} \cdot \frac{1\text{ mol Cr}}{3\cancel{F}} = \underline{0.0833\text{mol Cr}}$

17.56 $?A = \frac{50.0\cancel{\text{ml }O_2\text{(STP)}}}{3.00\cancel{\text{hr}}} \cdot \frac{4\cancel{F}}{22{,}400\cancel{\text{ml }O_2\text{(STP)}}} \cdot \frac{96{,}500\cancel{C}}{1\cancel{F}} \cdot \frac{1\cancel{\text{hr}}}{3600\cancel{s}} \cdot \frac{1\text{ A}}{1\cancel{C \cdot s^{-1}}} =$

$\underline{0.0798\text{ A}}$

17.57 $?M(OH^-) = ?\text{mol }OH^- = 1\cancel{\text{ltr soln}} \cdot \frac{0.250\cancel{A} \cdot 35.0\cancel{\text{min}}}{0.400\cancel{\text{ltr soln}}} \cdot \frac{1\cancel{C \cdot s^{-1}}}{1\cancel{A}} \cdot \frac{60\cancel{s}}{1\cancel{\text{min}}} \cdot$

$\frac{1\text{ mol }OH^-}{96{,}500\cancel{C}} = \underline{0.0136M(OH^-)}$ $\therefore$ $\underline{pH = 12.134}$

17.58 $2H_2O \rightleftharpoons H_2 + 2OH^- \; -2e^-$

$?A = \frac{15.5\cancel{\text{ml HCl}}}{25.0\cancel{\text{min}}} \cdot \frac{1\cancel{\text{min}}}{60\cancel{s}} \cdot \frac{0.250\cancel{\text{mol HCl}}}{10^3\cancel{\text{ml HCl}}} \cdot \frac{1\cancel{\text{mol }OH^-}}{1\cancel{\text{mol HCl}}} \cdot \frac{2\cancel{F}}{2\cancel{\text{mol }OH^-}} \cdot \frac{96{,}500\cancel{C}}{1\cancel{F}} \cdot \frac{1\text{ A}}{1\cancel{C \cdot s^{-1}}}$

$= \underline{0.249\text{ A}}$

17.59 (a) $Pb + SO_4^{2-} + Hg_2Cl_2 \longrightarrow 2Hg + 2Cl^- + PbSO_{4(s)}$

$+0.36V + 0.27V = \underline{0.63V = E^\circ}$

(17.59 continued) (b) $2Ag + 2Cl^- + Cu^{2+} \longrightarrow 2AgCl_{(s)} + Cu$ $E^\circ = 0.12V$

(c) $Mn + Cl_2 \longrightarrow Mn^{2+} + 2Cl^-$ $E^\circ = 2.39V$

(d) $2Al + 3Br_2 \longrightarrow 2Al^{3+} + 6Br^-$ $E^\circ = 2.76V$

17.60 (a) $E^\circ = 1.67V - 0.25V = 1.42V$ (b) $E^\circ = 1.69V - 1.33V = 0.36V$

(c) $E^\circ = 0.80V + 0.13V = 0.93V$ (d) $E^\circ = 1.36V - 1.28V = 0.08V$

(e) $E^\circ = 1.03V + 0.00V = 1.03V$

17.61 (a) $-0.38V$ (b) $-1.49V$ (c) $0.64V$ (d) $-0.16V$

(e) $-0.13V$

17.62 (a) $K_c = \{Ni^{2+}\}/\{Sn^{2+}\}$ $\Delta G = \Delta G^\circ + 2.303 \cdot RT \cdot \log(M.A.)$ AT EQUILIBRIUM $\Delta G = 0$ $(M.A.) = K_c$ $\therefore$ $\Delta G^\circ = -RT \cdot 2.303 \cdot \log K_c = -nFE^\circ$

$\log K_c = n E^\circ/0.0592$ $n = 2$ $E^\circ = 0.25V - 0.14V = 0.11V$ $K_c = 5 \cdot 10$

(b) $K_c = 10^{(2E^\circ/0.0592)}$ $E^\circ = 1.36V - 1.09V = 0.27V$ $K_c = 1 \cdot 10^9$

(c) $K_c = 10^{(E^\circ/0.0592)}$ $E^\circ = 0.80V - 0.77V = 0.03V$ $K_c = 3$

17.63 (a) $E^{\circ} = 0.77V + 0.14V = 0.91V$ $\log K_c = nE/0.0592$ $n = 2$

$\log K_c = 2 \cdot 0.91/0.0592 = 30.74$ $K_c = 6 \cdot 10^{30}$ (b) $K_c = 3 \cdot 10^{-12}$

(c) $K_c = 1 \cdot 10^{-72}$ (d) $K_c = 1 \cdot 10^{9}$ (e) $K_c = 9 \cdot 10^{47}$

17.64 (assume 25 C) $\Delta G^{\circ} = -nFE^{\circ} = -2.303 \cdot RT \cdot \log K_c$ $\log K_c = \dfrac{n E^{\circ}}{0.0592}$

(a) $n = 2$ $E^{\circ} = -2.76V + 2.38V = -0.38V$ $\log K_c = -12.84$ $K_c = 1 \cdot 10^{-13}$

(b) $E^{\circ} = -1.49V$ $n = 2$ $K_c = 5 \cdot 10^{-51}$

(c) $E^{\circ} = 0.64V$ $n = 2$ $K_c = 4 \cdot 10^{21}$

(d) $E^{\circ} = -0.16V$ $n = 30$ $K_c = 8 \cdot 10^{-82}$

(e) $E^{\circ} = -0.13V$ $n = 4$ $K_c = 2 \cdot 10^{-9}$

17.65 $\Delta G^{\circ} = -nFE^{\circ}$ (a) $\Delta G^{\circ} = -2\text{mol} \cdot \dfrac{96{,}500C}{1^{*}\text{mol } e^{-}} \cdot 0.91V \cdot \dfrac{10^{-3}kJ}{1^{*}V \cdot C} = -180\ kJ$

(b) $\Delta G^{\circ} = -2 \cdot 96{,}500 \cdot (-0.34) \cdot 10^{-3} = 66\ kJ$

(c) $\Delta G^{\circ} = -6 \cdot 96{,}500 \cdot (-0.71) \cdot 10^{-3} = 410\ kJ$

(d) $\Delta G^{\circ} = -2 \cdot 96{,}500 \cdot 0.27 \cdot 10^{-3} = -52\ 1K$

(e) $\Delta G^{\circ} = -2 \cdot 96{,}500 \cdot 1.42 \cdot 10^{-3} = -274\ kJ$

(FOR ANSWERS (a) - (d) the 2 s.f. of the voltage set a limit of 2 s.f. for the final answer.)

17.66 $\Delta G^\circ = -nFE^\circ$ (a) $E^\circ = 2.38V - 2.76V = \underline{-0.38V}$ $\underline{n = 2}$

$$\Delta G^\circ = -2\cancel{*mol\ e^-}\bullet\frac{96,500\cancel{C}}{1\cancel{*mol\ e^-}}\bullet(-0.38\cancel{V})\bullet\frac{1*\cancel{J}}{1*\cancel{V}\bullet\cancel{C}}\bullet\frac{10^{-3}*kcal}{4.184*\cancel{J}} = \underline{\underline{18\ kcal}}$$

(b) $E^\circ = -0.13V - 1.36V = \underline{-1.49V}$ $\underline{n = 2}$ $\Delta G^\circ = -2\bullet 96,500\bullet(-1.49)\bullet\frac{10^{-3}}{4.184}$ $= \underline{\underline{68.7\ kcal}}$

(c) $E^\circ = 2.00V - 1.36V = \underline{0.64V}$ $\underline{n = 2}$ $\Delta G^\circ = -2\bullet 96,500\bullet 0.64\bullet\frac{10^{-3}}{4.184} =$ $\underline{\underline{-30\ kcal}}$

(d) $E^\circ = 1.33V - 1.49V = \underline{-0.16V}$ $\underline{n = 30}$ $\Delta G^\circ = \cancel{-30}\bullet 96,500\bullet(-0.16)\bullet\frac{10^{-3}}{4.184} =$ $\underline{\underline{110\ kcal}}$

(e) $E^\circ = 1.23V - 1.36V = \underline{-0.13V}$ $\underline{n = 4}$ $\Delta G^\circ = -4\bullet 96,500\bullet(-0.13)\bullet\frac{10^{-3}}{4.184} =$ $\underline{\underline{12\ kcal}}$ (EXCEPT FOR (b) THE 2 s.f. OF THE VOLTAGE LIMIT THE FINAL ANSWERS TO 2 s.f.)

17.67 NERNST EQUATION: $E = E^\circ - \frac{0.0592}{n}\bullet\log(\text{mass action expression})$

(a) $\underline{\underline{E^\circ = 0.34V + 0.76V = 1.10V}}$ $\underline{\underline{E = 1.10V - 0.0592/2\ \log}}\frac{\{Zn^{2+}\}}{\{Cu^{2+}\}} = \underline{\underline{1.07V}}$

(b) $\underline{\underline{E^\circ = 0.25V - 0.14V = 0.11V}}$ $\underline{\underline{E = 0.11V - 0.0592/2\ \log(0.01)/(0.5) =}}$ $\underline{\underline{0.16V}}$

(c) $\underline{\underline{E^\circ = 3.05V + 2.87V = 5.92V}}$ $\underline{\underline{E = 5.92V - 0.0592/2\ \log}}\frac{Li^{+\ 2}\ F^{-\ 2}}{F_2} =$ $\underline{\underline{5.94V}}$

(d) $\underline{\underline{E^\circ = 0.76V + 0.0V = 0.76V}}$ $\underline{\underline{E = 0.76V - 0.0592/2\ \log}}\frac{(1\bullet 1)}{(0.01)^2} = \underline{\underline{0.64V}}$

(17.67 continued) (e) $E^\circ = 0.44V$ $E = 0.46V$

17.68 (a) $E^\circ = 1.67V - 0.25V = 1.42V$ $E = E^\circ - 0.0592/n \log(M.A.)$
$E = 1.42V - 0.0592/6 \log \frac{(0.020)^2}{(0.80)^3} = 1.45V$ $\Delta G = -nFE = -6 \bullet 96500 \bullet 1.45 \bullet 10^{-3} = 840\ kJ$

(b) $E^\circ = -0.14V + 0.25V = 0.11V$ $E = E^\circ - 0.0592/2 \log(0.010)/(1.10) = 0.17V$ $\Delta G = -2 \bullet 96500 \bullet 0.17 \bullet 10^{-3} = -33\ kJ$

(c) $E^\circ = 0.52V + 0.76V = 1.28V$ $E = 1.28V - 0.0592/2 \log(0.010)/(0.050)^2 = 1.26V$ $\Delta G = -2 \bullet 96500 \bullet 1.26 \bullet 10^{-3} = -243\ kJ$

17.69 $E = E^\circ - 0.0592/n \log(M.A.)$ $E = (0.14V - 0.13V) - 0.0592/2 \bullet \log(1.50)/(0.010) = -0.03V$ (a)

(b) $E = (0.76V - 0.74V) - 0.0592/6 \log(0.020)^3/(0.010)^2 = 0.03V$

(c) $E = (1.69V - 0.34V) - 0.0592/2 \log(0.0010)/(0.010)(0.10)^4 = 1.26V$

17.70 $E = E^\circ - 0.0592/n \log(M.A.)$ $E = 0 - 0.0592/2 \log(0.0010/0.10) = 0.059V$

17.71 $E = E^\circ - 0.0592/n\ \log(M.A.)$ $E = 0 - 0.0592/3\ \log(0.0020/0.10$
$= \underline{\underline{0.035V}}$

17.72 $E = E^\circ - 0.0592/n\ \log(M.A.)$ $\frac{1}{2}H_2 + Ag^+ \longrightarrow H^+ + Ag$
$E = (0V + 0.80V) - 0.0592/1\ \log(\{H^+\}/\{Ag^+\})$ $K_{sp} = 5\bullet10^{-13} = \{Ag^+\}\{Br^-\}$
$\underline{\{Ag^+\} = 5\bullet10^{-13}/0.010 = 5\bullet10^{-11}}$ $E = 0.80V - 0.0592\bullet\log(1/5\bullet10^{-11}) =$
$\underline{\underline{0.19V}}$

17.73 $E = 0.34V - 0.0296\ \log(1/(2\bullet10^{-4}) = \underline{\underline{0.23V}}$

17.74 $E = E^\circ - 0.0592/n\ \log\{Pb^{2+}\}$ $\log\{Pb^{2+}\} = (E^\circ - E)\bullet2/0.0592$
$\log\{Pb^{2+}\} = (0.13V - 0.51V)\bullet2/0.0592 = \overline{-12.84}$ $\underline{\{Pb^{2+}\} = 1\bullet10^{-13}}$
$\underline{\underline{K_{sp} = \{Pb^{2+}\}\{CrO_4^{2-}\} = (1\bullet10^{-13})(0.10) = 1\bullet10^{-14}}}$

17.75 $Mg_{(s)} + 2Ag^+ \longrightarrow Mg^{2+} + 2Ag_{(s)}$ $E^\circ = 2.38V + 0.80V =$
$\underline{3.18V}$ $E = E^\circ - 0.0592/n\ \log\frac{\{Mg^{2+}\}}{\{Ag^+\}^2}$ (a) $E = 3.18V - 0.0296\ \log\frac{(0.100)}{(0.100)^2}$

(17.75 continued) $\underline{\underline{E = 3.15V}}$

(b) $?\text{mol } Ag^+ = 1.00^\bullet \text{g Ag} \bullet \frac{1^*\text{mol } Ag^+}{107.9\text{g } Ag^+} = \underline{9.268 \bullet 10^{-3} \text{mol } Ag^+}$

$?\text{mol } Mg^{2+} = 9.268 \bullet 10^{-3} \text{mol } Ag^+ \bullet \frac{1^*\text{mol } Mg^{2+}}{2^*\text{mol } Ag^+} = \underline{4.634 \bullet 10^{-3} \text{mol } Mg^{2+}}$

$M_f = M_i + \Delta M \qquad ?M_f(Ag^+) = 0.100^\bullet M_i(Ag^+) - \frac{9.268 \bullet 10^{-3} \text{mol } Ag^+}{0.200 \text{ltr } Ag^+ \text{soln}} = \underline{0.054M_f}$

$?M_f(Mg^{2+}) = 0.100^\bullet M_i(Mg^{2+}) + \frac{4.634 \bullet 10^{-3} \text{mol } Mg^{2+}}{0.250 \text{ltr } Mg^{2+} \text{soln}} = \underline{0.119M_f(Mg^{2+})}$

$E = 3.18V - 0.0296 \log \frac{(0.119)}{(0.054)^2} = \underline{\underline{3.13V}}$

(c) $?\text{mol } Mg^{2+} = 0.080^\bullet \text{g Mg} \bullet \frac{1^*\text{mol } Mg^{2+}}{24.31\text{g Mg}} = \underline{3.29 \bullet 10^{-3} \text{mol } Mg^{2+}}$

$?\text{mol } Ag^+ = 3.29 \bullet 10^{-3} \text{mol } Mg^{2+} \bullet \frac{2^*\text{mol } Ag^+}{1^*\text{mol } Mg^{2+}} = \underline{6.58 \bullet 10^{-3} \text{mol } Ag^+}$

$?M_f Ag^+ = 0.100^\bullet M_i Ag^+ - \frac{6.58 \bullet 10^{-3} \text{mol } Ag^+}{0.200 \text{ltr } Ag^+ \text{soln}} = \underline{0.067M_f Ag^+}$

$?M_f Mg^{2+} = 0.100^\bullet M_i Mg^{2+} + \frac{3.29 \bullet 10^{-3} \text{mol } Mg^{2+}}{0.250 \text{ltr } Mg^{2+} \text{soln}} = \underline{0.113M_f Mg^{2+}}$

$E = 3.18V - 0.0296 \log \frac{(0.113)}{(0.067)^2} = \underline{\underline{3.14V}}$

17.76 $\frac{\{H^+\}^2\{S^{2-}\}}{\{H_2S\}} = 1.1\cdot10^{-21} = \frac{(0.0010)^2\{S^{2-}\}}{(0.10)}$ $\underline{\{S^{2-}\} = 1.1\cdot10^{-16}}$

$\{Ag^+\}^2\{S^{2-}\} = 2\cdot10^{-49} = \{Ag^+\}^2(1.1\cdot10^{-16})$ $\underline{\{Ag^+\} = 4.26\cdot10^{-17}}$

$E = E^\circ - 0.0592/n\ \log(1/\{Ag^+\}) = 0.80V - 0.0592\ \log(2.35\cdot10^{16}) = \underline{\underline{-0.17V}}$

17.77

	(I)	(STP)
P	767-27 torr	760 torr
V	288 ml	? ml
T	300 K	273 K

$?ml(STP) = 288ml(I)\cdot\frac{740\cancel{torr}\cdot273\cancel{K}}{760\cancel{torr}\cdot300\cancel{K}} = \underline{255ml(STP)}$

$?C = 1^*\cancel{e^-}\cdot\frac{1.22\,Ⓒ\cdot\cancel{s^{-1}}\cdot30.0\cancel{min}}{255\cancel{ml\ H_2\ (STP)}}\cdot\frac{60^*\cancel{s}}{1^*\cancel{min}}\cdot\frac{11,200\cancel{ml\ H_2\ (STP)}}{6.022\cdot10^{23}\cancel{e^-}} = \underline{\underline{1.60\cdot10^{-19}C\text{ (charge on an electron in coulombs)}}}$

17.78 $?hr = 25.0\cancel{g\ Pb}\cdot\frac{2^*\cdot96,500\cancel{C}}{207.2\cancel{g\ Pb}}\cdot\frac{1.5^*\cancel{V}}{25^*\cancel{W}}\cdot\frac{1^*\cancel{W}}{1^*\cancel{V}\cdot\cancel{C}\cdot\cancel{s^{-1}}}\cdot\frac{1^*ⓗⓡ}{3600\cancel{s}} = \underline{\underline{0.39\ hr}}$

17.79 $?A = 1^*\cancel{m^2}\ \cancel{Cr(plate)}\cdot\frac{7.19\cancel{g\ Cr}}{1^*\cancel{cm^3}\ \cancel{Cr}}\cdot\frac{6^*\cdot96,500Ⓒ}{52.00\cancel{g\ Cr}}\cdot\frac{0.050\cancel{mm}}{\cancel{(plate)}\cdot25\cancel{min}}\cdot\frac{1^*\cancel{cm}}{10^*\cancel{mm}}\cdot\frac{(100^*\cancel{cm})^2}{1^*\cancel{m^2}}\cdot\frac{1^*\cancel{min}}{60^*Ⓢ} = \underline{\underline{3\cdot10^3\ A}}$ (only 1 s.f. because of 1 m^2)

17.80 $\Delta G^\circ = \Delta H^\circ - T\Delta S^\circ = -242\text{kJ} - 383\cancel{K} \bullet (188.7 - 130.6 - 205.0/2) \bullet 10^{-3}\text{kJ}/\cancel{K}$

$= \underline{-225 \text{ kJ}}$ $\quad ?\text{g } H_2 = 1^*\cancel{s} \bullet 1.0 \bullet 10^3 \cancel{W} \bullet \frac{1^*\cancel{J(net)}}{1^*\cancel{W \bullet s}} \bullet \frac{100^*\cancel{J(gross)}}{70^*\cancel{J(net)}} \bullet \frac{2.016\text{g } H_2}{225 \bullet 10^3 \cancel{J(gross)}}$

$= \underline{\underline{0.013\text{g } H_2 \text{ (mass reacting each second)}}}$ $\quad \underline{\underline{(0.10\text{g } O_2)}}$

17.81 $\frac{\{H^+\}\{C_2H_3O_2^-\}}{\{HC_2H_3O_2\}} = 1.8 \bullet 10^{-5} = \frac{\{H^+\}^2}{0.10}$ $\quad \underline{\{H^+\}^2 = 1.8 \bullet 10^{-6}}$

$2H^+ + Fe \longrightarrow H_2 + Fe^{2+}$ $\quad E = E^\circ - (0.0592/n) \bullet \log \frac{\{Fe^{2+}\}}{\{H^+\}^2}$

$E = (0.0V + 0.44V) - (0.0296) \bullet \log(0.10)/(1.8 \bullet 10^{-6}) = \underline{\underline{0.30 \text{ V}}}$

17.82 $?\text{kJ} = 5.00 \bullet \cancel{min} \bullet 110\cancel{V} \bullet 1.00\cancel{A} \bullet \frac{1^*\cancel{W}}{1^*\cancel{V \bullet A}} \bullet \frac{1^*\text{kJ}}{10^3 {}^*\cancel{W \bullet s}} \bullet \frac{60^*\cancel{s}}{1^*\cancel{min}} = \underline{\underline{33.0\text{kJ}}}$

17.83 $C_{12}H_{26(l)} + 18\frac{1}{2}\, O_{2(g)} \longrightarrow 12CO_{2(g)} + 13H_2O_{(g)}$

$\Delta H^\circ_{comb} = 12 \bullet \Delta H^\circ_f(CO_2) + 13 \bullet \Delta H^\circ_f(H_2O_{(g)}) - \Delta H^\circ_f(C_{12}H_{26(l)}) = (12 \bullet (-394) +$

$13 \bullet (-242) - (291))\text{kJ} = \underline{-8165\text{kJ}}$ $\quad ?\text{ltr}(C_{12}H_{26(l)}) = 1.0 \bullet \cancel{kW \bullet hr} \bullet \frac{1^*\cancel{kJ}}{1^*\cancel{kW \bullet s}} \bullet$

$\frac{3600^*\cancel{s}}{1^*\cancel{hr}} \bullet \frac{1^*\cancel{mol(C_{12}H_{26})}}{8165\cancel{kJ}} \bullet \frac{170\cancel{g}(C_{12}H_{26})}{1^*\cancel{mol\ C_{12}H_{26}}} \bullet \frac{10^{-3} {}^*\text{ltr}}{0.74 \bullet \cancel{g}} \bullet \frac{100 \text{ gross}}{30 \text{ net}} = \underline{\underline{0.341\text{ltr}(C_{12}H_{26})}}$

17.84 $?\text{min} = 0.500\,\cancel{\text{ltr soln}} \bullet \frac{0.270 * \cancel{\text{mol } Cr_2(SO_4)_3}}{1 * \cancel{\text{ltr soln}}} \bullet \frac{6 \bullet 96{,}500 \cancel{C}}{1 * \cancel{\text{mol } Cr_2(SO_4)_3}} \bullet \frac{1 * \cancel{s}}{3.00 \cancel{C}} \bullet \frac{1 * \text{min}}{60 * \cancel{s}} = \underline{\underline{434 \text{ min}}}$

17.85 $E^\circ = E^\circ(Mn^{2+}/Mn^{7+}) + E^\circ(Cr^{6+}/Cr^{3+}) = -1.49V + 1.33V = \underline{-0.16V}$

$\Delta G = \Delta G^\circ + 2.303RT \log \frac{\{MnO_4^-\}^6 \{Cr^{3+}\}^{10}}{\{Mn^{2+}\}^6 \{Cr_2O_7^{2-}\}^5 \{H^+\}^{22}}$ $\Delta G^\circ = -nFE^\circ = -30 \cancel{\text{mol } e^-} \bullet$

$\frac{96{,}500 \cancel{C}}{1 * \cancel{\text{mol } e^-}} \bullet (-0.16 \cancel{V}) \bullet \frac{10^{-3} * \text{kJ}}{1 * \cancel{V} \bullet \cancel{C}} = \underline{463.2 \text{ kJ}}$ $\Delta G = 463.2\text{kJ} + 2.303 \bullet 8.314 \cancel{J} \bullet \cancel{K}^{-1}$

$\bullet 298 \cancel{K} \bullet \frac{1 * \text{kJ}}{10^3 * \cancel{J}} \bullet \log \frac{(0.0010)^6 (0.0010)^{10}}{(0.10)^6 (0.010)^5 (1.0 \bullet 10^{-6})^{22}} = 463.2\text{kJ} + 5.71\text{kJ} \bullet \log 10^{1(}$

$= \underline{\underline{1.0 \bullet 10^3 \text{ kJ}}}$ Since the reaction as written involves free energy increase, the spontaneous reaction would be the reverse reaction. <u>(reaction to the left)</u>

17.86 ANODE REACTION: $(?M)Cl^- + Ag \longrightarrow AgCl + 1e^-$
CATHODE REACTION: $AgCl + 1e^- \longrightarrow Ag + (1M)Cl^-$

$(?M)Cl^- \longrightarrow (1M)Cl^-$ (a concentration cell)

$E = 0.0435V$ $\underline{E^\circ = 0}$ $E = E^\circ - (0.0592/1) \bullet \log(1)/\{Cl^-\}_A$

$-\log(1/\{Cl^-\}_A = 0.0435V/(0.0592V)$ $\underline{\underline{\{Cl^-\}_A = 5.43M}}$

17.87 What is the S^{2-} concentration in a solution saturated with H_2S?

$$\frac{\{H^+\}\{HS^-\}}{\{H_2S\}} = 1.1 \cdot 10^{-7} \qquad \frac{\{H^+\}\{S^{2-}\}}{\{HS^-\}} = 1.0 \cdot 10^{-14}$$

Since $\{H^+\} \approx \{HS^-\}$, the $\underline{\{S^{2-}\} = 1.0 \cdot 10^{-14}}$

$$E = E^\circ - (0.0592/n) \cdot \log \frac{\{Zn^{2+}\}}{\{Cu^{2+}\}} \qquad \log\{Zn^{2+}\} - \log\{Cu^{2+}\} = \frac{E - E^\circ}{-(0.0592/2)}$$

$$\log\{Cu^{2+}\} = \frac{(0.67V - 1.10V)}{(0.0296)} + \log(1.0) = -14.53 \qquad \underline{\{Cu^{2+}\} = 3.0 \cdot 10^{-15}}$$

$K_{sp} = \{Cu^{2+}\}\{S^{2-}\} = (3.0 \cdot 10^{-15})(1.0 \cdot 10^{-14}) = \underline{\underline{3.0 \cdot 10^{-29}}}$ BUT $\underline{K_{sp} = 8.5 \cdot 10^{-36}}$

(table 16.1) Why is the K_{sp} which we calculated more than a million times larger? The clue is the problem statement that, "The solution -- formed was not buffered." As S^{2-} is consumed by the precipitation of CuS, more H_2S ionizes to supply additional S^{2-}. FOR EACH MOLE OF S^{2-} PRODUCED, TWO MOLES OF H^+ ARE ALSO SUPPLIED. Since the solution is not buffered the concentration of H^+ increases to about 0.20M

$$\frac{\{H^+\}^2\{S^{2-}\}}{\{H_2S\}} = 1.1 \cdot 10^{-21} = \frac{(0.20)^2\{S^{2-}\}}{(0.10)} \qquad \{S^{2-}\} = 2.8 \cdot 10^{-21}$$

using this value the $\underline{\underline{K_{sp} = 8.4 \cdot 10^{-36}}}$ (This value is in close agreement with the $\underline{8.5 \cdot 10^{-36}}$ found in table 16.1.)

PROBLEM SOLUTIONS FOR CHAPTER 18 of BRADY & HUMISTON 3rd Ed.

18.75 $Na^+_{(g)} = Na^+_{(aq)}$ $\Delta H° = ?$ (1) $Na_{(s)} = Na^+_{(aq)}$ $\Delta H°_1 = -57.3$

(2) $Na_{(g)} = Na_{(s)}$ $\Delta H°_2 = -25.98$ (3) $Na^+_{(g)} = Na_{(g)}$ $\Delta H°_3 = -118.0$

$\Delta H° = \Delta H°_1 + \Delta H°_2 + \Delta H°_3 =$ $\underline{\underline{-201 \text{ kcal mol}^{-1}}}$

18.76 $Na_{(s)} = Na^{2+}_{(aq)} + 2\ e^-$ (overall process) $\Delta H°$ =(negative)

$Na_{(s)} = Na_{(g)}$ $\Delta H°_{atomization} = 25.98 \text{ kcal mol}^{-1}$

$Na_{(g)} = Na^+_{(g)} + 1\ e^-$ $I.E._{(1)} = 118.0 \text{ kcal mol}^{-1}$

$Na^+_{(g)} = Na^{2+}_{(g)} + 1\ e^-$ $I.E._{(2)} = 1091 \text{ kcal mol}^{-1}$

$Na^{2+}_{(g)} = Na^{2+}_{(aq)}$ $\Delta H°_{hydration} = ?$ (<u>must be a larger negative number than</u>

<u>$-1235 \text{ kcal mol}^{-1}$</u>)

18.77 $ZnO_{(s)} = Zn + 1/2\ O_2$ $\Delta H° = +83.2 \text{ kcal mol}^{-1}$, $\Delta S° =$ (10.0+

(cont.)

(18.77 continued)

1/2 x 49.0 - 10.4)cal $(mol \bullet K)^{-1}$ $\Delta G^\circ_T = \Delta H^\circ - T\Delta S^\circ = 83.2$ kcal$-T(24.1 \bullet 10^{-3}$ kcal $K^{-1}) = 0$ $\therefore$ T = 3450 K (3180° C)

18.78 $\Delta G^\circ = -2.303RT \log K_p$ $\Delta G^\circ = \Delta H^\circ - T\Delta S^\circ$ IF: $K_p = 1$ $\therefore$ $\Delta G^\circ = 0$

$\therefore$ $\Delta H^\circ = T\Delta S^\circ$ $\Delta H^\circ = 155$ kJ mol^{-1} $\Delta S^\circ = S^\circ(Cu_{(s)}) + 1/2\ S^\circ(O_{2(g)}) -$ $S^\circ(CuO_{(s)})$ $T = \Delta H^\circ/\Delta S^\circ$ $\Delta S^\circ = 33.3\ J(mol \bullet K)^{-1} + 1/2 \times 205.0\ J(mol \bullet K)^{-1}$ $- 43.5\ J(mol \bullet K)^{-1}$ $\Delta S^\circ = 92.3\ J(mol \bullet K)^{-1}$ $\therefore$ $T = 1.55 \bullet 10^5 J \bullet mol^{-1}/(92.3$ $J(mol \bullet K)^{-1})^{-1}$ = 1680 K (3 s.f.)

18.79 $\Delta G^\circ = -RT \ln K_p$ $\Delta G^\circ = \Delta H^\circ - T\Delta S^\circ = 180.3$kcal $- T(61.65 \frac{cal}{K})$

$\Delta G^\circ_{373} = 157.3$kcal $\Delta G^\circ_{773} = 132.6$kcal $\Delta G^\circ_{2273} = 40.2$kcal

$\ln K_p = -\Delta G^\circ/RT$ $K_p = e^{-\Delta G^\circ/RT}$ $K_p(373) = 6.71 \bullet 10^{-93}$

$K_p(773) = 3.21 \bullet 10^{-38}$ $K_p(2273) = 1.36 \bullet 10^{-4}$

18.80

$K_{(s)}$	$\rightarrow$	$K^+_{(aq)}$	+ $1e^-$	$\Delta H^\circ = ?$
$K_{(s)}$	$\rightarrow$	$K_{(g)}$		$\Delta H^\circ = 90.0\,kJ$
$K_{(g)}$	$\rightarrow$	$K^+_{(g)}$	+ $1e^-$	$\Delta H^\circ = 418\ kJ$
$K^+_{(g)}$	$\rightarrow$	$K^+_{(aq)}$		$\Delta H^\circ = -759\ kJ$
				$\Delta H^\circ = -251\ kJ$

$$\Delta G^\circ = -nFE^\circ = \Delta H^\circ - T\Delta S^\circ$$

$$\Delta S^\circ = \frac{-nFE^\circ - \Delta H^\circ}{-T} = (1 \bullet 96{,}500\,C \bullet 2.92\,V \bullet \frac{1*J}{1*V \bullet C})/298\,K + (-251 \bullet 10^3\,J)/298\,K =$$

$103\ J \bullet K^{-1}$

PROBLEM SOLUTIONS FOR CHAPTER 19 OF BRADY & HUMISTON 3rd Ed.

19.101 $\Delta H_{rctn(1)} = 8*\text{mol } CO_{2(g)}(-394 \text{ kJ}\cdot\text{mol}^{-1}) + 9*\text{mol} H_2O_{(g)}$ x

$(-242 \text{ kJ}\cdot\text{mol}^{-1}) - 1*\text{mol } C_8H_{18(l)}(-255.1 \text{ kJ}\cdot\text{mol}^{-1}) - 0 = \underline{\underline{-5075 \text{ kJ}}}$

$\Delta H_{rctn(2)} = 8*\text{mol } CO_{2(g)}(-394 \text{ kJ}\cdot\text{mol}^{-1}) + 9*\text{mol } H_2O_{(g)}(-242 \text{ kJ}\cdot\text{mol}^{-1}) +$

$0 - 1*\text{mol } C_8H_{18(l)}(-255.1 \text{ kJ}\cdot\text{mol}^{-1}) - 25*\text{mol } N_2O_{(g)}(81.5 \text{ kJ}\cdot\text{mol}^{-1}) =$

$\underline{\underline{-7112 \text{ kJ}}}$

PROBLEM SOLUTIONS FOR CHAPTER 21 OF BRADY & HUMISTON 3rd Ed.

21.93 $K_{sp} = [Hg^+][Cl^-]^2 = 2.2 \cdot 10^{-18},\ 4.4 \cdot 10^{-18},\ 1.1 \cdot 10^{-17},\ 2.2 \cdot 10^{-17}$

NOT CONSTANT

$K_{sp} = [Hg_2^{2+}][Cl^-]^2 = 1.1 \cdot 10^{-18},\ 1.1 \cdot 10^{-18},\ 1.1 \cdot 10^{-18},\ 1.1 \cdot 10^{-18}$

CONSTANT VALUE $\therefore$ correct formula = Hg_2^{2+}

PROBLEM SOLUTIONS FOR CHAPTER 24 of BRADY & HUMISTON 3rd Ed.

24.26 (a) $k = \dfrac{0.693}{t_{1/2}}$ $kt_{1/2} = 0.693$ $\therefore$ kt = 0.693 x no. of

half-life periods $\log\, (^{60}Co)_0/(^{60}Co) = \dfrac{1 \times 0.693}{2.303} = \log \dfrac{1.00g\,^{60}Co}{(^{60}Co)_a} =$

$0.3009 \quad \dfrac{1.00g\,^{60}Co}{(^{60}Co)_a} = 0.500 \quad \therefore \quad \underline{\underline{(^{60}Co)_a = 0.500\ g}}$

(b) $\log\, (^{60}Co)_0/(^{60}Co)_b = \dfrac{3 \times 0.693}{2.303} = 0.9027 \quad \underline{\underline{(^{60}Co)_b = 0.125\ g}}$

$\underline{\underline{(^{60}Co)_c = 0.031\ g}}$ (AN ALTERNATIVE METHOD IS SHOWN BELOW)

When "n" is the number of half-life periods, the fraction of the original sample which remains = $1/2^n$.

(a) $\underline{\underline{1.00g \times 1/2^1 = 0.500\ g}}$ (b) $\underline{\underline{\times 1/2^3 = 0.125\ g}}$ (c) $\underline{\underline{\times 1/2^5 = 0.031\ g}}$

24.27 (see 24.26) (a) $\log \dfrac{8.00g}{X_a} = \dfrac{240}{120} \times \dfrac{0.693}{2.303} = 0.6018 \quad \underline{\underline{X_a = 2.00g}}$ (no. of half-life periods)

(24.27 continued) (b) (4 half-life periods) $X_b = 0.50$ g

(c) (8) $X_c = 0.031$ g

24.28

$$t_{1/2} = \frac{0.693}{k} = \frac{0.693}{4.23 \cdot 10^{-3}\ \text{days}^{-1}} = \underline{\underline{164\ \text{days}}}$$

24.29

$$t_{1/2} = \frac{0.693}{k} = \underline{\underline{3.01 \cdot 10^{5}\text{yr}}}$$

24.30

$$\log \frac{(^{51}\text{Cr})_0}{(^{51}\text{Cr})_t} = \frac{kt}{2.303} = \log 2 \quad (\text{when } t = 27.72 \text{ days})$$

$$\therefore\ k = \frac{2.303 \cdot \log 22}{27.72 \text{ days}} \cdot \frac{1^{*}\text{day}}{24^{*} \cdot 60^{*} \cdot 60^{*}\text{s}} = \underline{\underline{2.895 \cdot 10^{-7}\text{s}^{-1}}}$$

24.31

$$k = \frac{0.693}{t_{1/2}} = \underline{\underline{0.00147\ \text{day}^{-1}}}$$

24.32 mol ^{40}Ar formed = mol ^{40}K decayed = $1.15 \cdot 10^{-5}$mol

$$K = \frac{0.693}{t_{1/2}} = \frac{0.693}{1.3 \cdot 10^{9}yr} = \underline{5.3 \cdot 10^{-10} yr^{-1}} \qquad t = \frac{2.303}{k} \cdot \log \frac{\{A\}_0}{\{A\}_t}$$

$$\underline{\{A\}_0 = (2.07 \cdot 10^{-5} + 1.15 \cdot 10^{-5}) \text{mol } ^{40}K} \qquad \underline{\{A\}_t = 2.07 \cdot 10^{-5} \text{ mol } ^{40}K}$$

$$t = \frac{2.303}{5.3 \cdot 10^{-10} yr^{-1}} \cdot \log \frac{3.22 \cdot 10^{-5}}{2.07 \cdot 10^{-5}} = \underline{\underline{8.3 \cdot 10^{8} yr}}$$

24.33 Since three half-life periods would reduce the ^{14}C to one eight the initial value, the age of the wood equals $3 \cdot 5770$yr.(17,300)

24.34 $E = mc^2 \qquad E = 2 \cdot 9.1096 \cdot 10^{-31} kg(2.9979 \cdot 10^{8} m \cdot s^{-1})^2 =$

$\underline{\underline{1.6374 \cdot 10^{-13} J \quad (1^* kg \cdot m^2 \cdot s^{-2} = 1^* J)}}$

24.35 $\Delta m(^{7}Li) = (3 \cdot 1.007277 + 4 \cdot 1.008665 + 3 \cdot 0.0005486) - (7.01600) =$

$0.042137 amu \qquad E = \Delta m \cdot c^2 = 0.042137 \not{g} \cdot mol^{-1} \cdot (2.9979 \cdot 10^{8} m \cdot s^{-1})^2 \cdot \frac{1^* kg}{10^{3} {}^* \not{g}} =$

$\underline{\underline{3.7870 \cdot 10^{12} J \cdot mol^{-1} \quad (3.7870 \cdot 10^{9} kJ \cdot mol^{-1}) \quad (39.250 \text{ MeV})}}$

$\underline{\underline{\text{for: } ^{19}F \ 1.4261 \cdot 10^{10} kJ \cdot mol^{-1}, \ 147.80 MeV \quad ^{14}N \ 1.0098 \cdot 10^{10} kJ, \ 104.65 MeV}}$

24.36 $^{56}Fe = 26p \bullet 1.007277 amu/p + 30n \bullet 1.008665 amu/n + 2e \bullet 5.4859 \bullet 10^{-4} amu/e$

$= \underline{56.463415 amu}$ Since the actual atomic mass = 55.9349amu, the mass defect = $\underline{0.5285 amu}$. What is the binding energy per nucleon?

$$?MeV = 1^{*}nucleon \bullet \frac{0.5285 amu}{56^{*}nucleon} \bullet \frac{931 MeV}{1^{*}amu} = \underline{\underline{8.79 MeV}}$$

Since this is the largest value of binding energy per nucleon (the highest point on the curve in figure 24.12) neither fission nor fusion of iron 56 can yield energy.

24.37 $\Delta m = 2 \bullet 2.014102 - 4.002603 = 0.025601$ $\quad E = \Delta m \bullet c^2$

$E = (0.025601 amu)(2.9979 \bullet 10^8 m \bullet s^{-1})^2 \bullet 6.022 \bullet 10^{23} mol^{-1} \bullet 1^{*}kg/6.022 \bullet 10^{26} amu \bullet 1^{*}kJ/10^3{}^{*}kg \bullet m^2 \bullet s^{-2} = \underline{\underline{2.3009 \bullet 10^9 kJ \bullet mol^{-1}}}$ (This answer was not limited to 4 s.f. because the Avogadro number was expressed to only four since the 6.022 cancelled exactly in numerator and denominator.)

24.38

$$k = \frac{1}{t} \bullet \ln\frac{\{A\}_o}{\{A\}_t}$$

$$k = \frac{1}{96.0 hr} \bullet \ln\frac{4720}{2560}$$

$$\underline{\underline{k = 6.37 \bullet 10^{-3} hr^{-1}}}$$

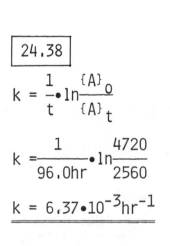

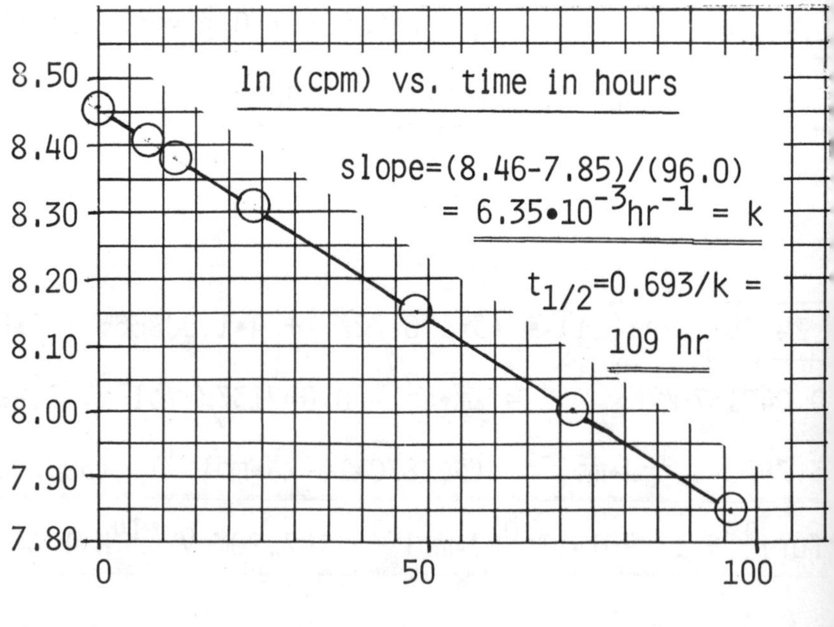

24.39 (10ml CH_3OH) (580 $cpm \cdot g^{-1}$) (29 $cpm \cdot g^{-1}$)

$$?ml(cool.) = 10ml\ CH_3OH \cdot \frac{0.792g\ CH_3OH}{1^*ml\ CH_3OH} \cdot \frac{580\ cpm}{1^*g\ CH_3OH} \cdot \frac{1^*ml(cool.)}{0.884g(cool.)} \cdot \frac{1^*g(cool.)}{29^*cpm}$$

$= 180\ ml(cool.)$ (2 s.f.)

24.40 $$?mol\ Cr = 165\ cpm \cdot \frac{1^*g\ K_2Cr_2O_7}{843\ cpm} \cdot \frac{2^*mol\ Cr}{294g\ K_2Cr_2O_7} = 1.33 \cdot 10^{-3} mol\ Cr$$

$$?mol\ C_2O_4^{2-} = 83\ cpm \cdot \frac{1^*g\ H_2C_2O_4}{345\ cpm} \cdot \frac{1^*mol\ C_2O_4^{2-}}{90.0g\ H_2C_2O_4} = 2.67 \cdot 10^{-3} mol\ C_2O_4^{2-}$$

$\therefore$ There are two oxalate ions bound to each Cr(III) in the complex ion.

24.41 (a) $2.3009 \cdot 10^9 kJ \cdot mol^{-1}$ (see 24.37)

(b) $2 \cdot 12.00000^\dagger - 23.98504 = 0.01496$ $\therefore$ $E = 1.345 \cdot 10^9 kJ \cdot mol^{-1}$

Reaction (a) produces more energy per mole of product.

On the basis of energy produced per gram of reactants, reaction (a) wins be a larger margin, slightly more than 10:1.

† Since carbon 12 is used as the basis of the atomic mass scale, the mass of this isotope is exactly 12.

24.42 $C_8H_{18(l)} + 12\frac{1}{2}O_{2(g)} \longrightarrow 8CO_{2(g)} + 9H_2O_{(l)}$

$$8\cdot\Delta H_f(CO_{2(g)}) + 9\cdot\Delta H_f(H_2O_{(l)}) - \Delta H_f(C_8H_{18(l)}) = \Delta H_{comb.}(C_8H_{18(l)})$$

$$\Delta H_{comb.}(C_8H_{18(l)}) = 8\cdot(-394)\text{kJ} + 9\cdot(-286)\text{kJ} - (-208.4)\text{kJ} = \underline{-5518\text{ kJ}}$$

$$?\text{gal } C_8H_{18(l)} = 1^*\text{mol } {}^4_2He\cdot\frac{2.3009\cdot10^9\text{kJ}}{1^*\text{mol } {}^4_2He}\cdot\frac{1^*\text{mol } C_8H_{18}}{5518\text{ kJ}}\cdot\frac{114\text{g } C_8H_{18}}{1^*\text{mol } C_8H_{18}}\cdot\frac{1\text{ltr}C_8H_{18}}{703\text{g}C_8H_{18}}$$

$$\cdot\frac{1^*\text{gal } C_8H_{18}}{3.79\text{ltr } C_8H_{18}} = \underline{\underline{1.78\cdot10^4\text{ gal } C_8H_{18(l)}}}$$